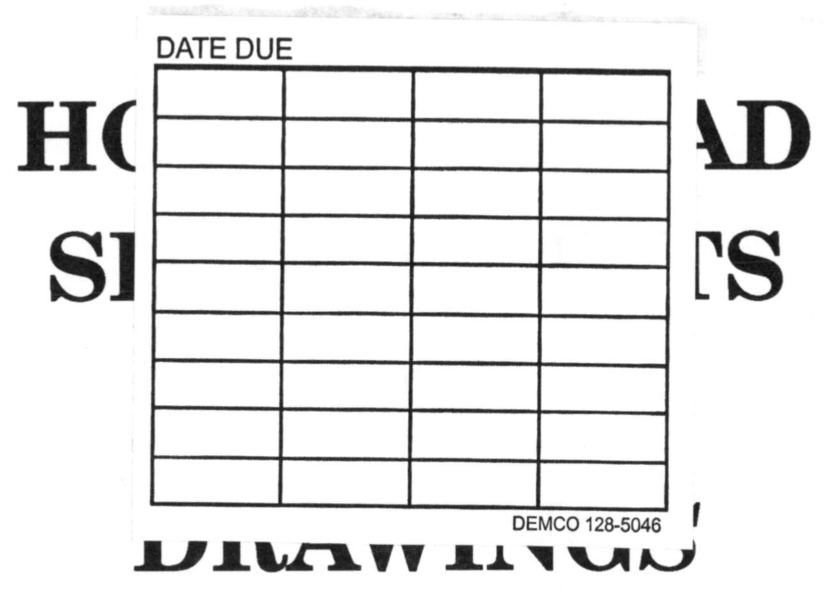

By

William E. Hardman

NATIONAL TOOLING AND MACHINING ASSOCIATION
Textbook Series
9300 Livingston Road
Fort Washington, Maryland 20744

REVISED 1997

NATIONAL TOOLING AND MACHINING ASSOCIATION
9300 Livingston Road
Fort Washington, Maryland 20744

301-248-6200
301-248-2755 (fax)
Publications only 1-800-832-7753

NTMA Catalog # 5003

ISBN-0-910399-01-8

Library of Congress
Catalog Card Number
82-61482

PREFACE

How to Read Shop Prints and Drawings is written for secondary, post-secondary and apprentice level machine trades programs which require the student to gain proficiency in machine shop print reading applications. It is equally well suited for metalworking manufacturing personnel who must interpret engineering drawings. These include machine operators, quality control inspectors, NC programmers, shop supervisors and many engineering managers. The content is based on technical guidance provided by machine shop owners among the 3,000 member companies of the National Tooling & Machining Association.

To undertake a career in precision metalworking and all associated metalworking fields, the student must learn to understand a new language. That is, to read the blueprint, the universal form of communication in machine shops and manufacturing plants. The NTMA compression-of-experience concept teaches the student the fundamentals of this new language in a rapid, but extremely practical way.

A fundamental part of NTMA's learning program is the use of actual shop prints taken from industry. The prints have been selected as representative of work that journeymen toolmakers, machinists or other manufacturing personnel, are likely to encounter in diversified job shop operations. For students in machine tool technology and apprentice programs, it is recommmended that this course be taught concurrently with instruction in machine shop theory and mathematics, as this course will provide the opportunity to put much of the related studies to practical use. The use of NTMA's Metalworking Training System (MTS) makes this easier to accomplish and the MTS meets National Institute for Metalworking Skills Standards.

Upon the successful completion of this program, the student will have gained proficiency in reading engineering drawings and will have acquired a foundation in the basic communications skills common to all metalworking manufacturers. In short, the student will have taken an important step in preparing for a metalworking career, whether it be hands-on operation or production management.

Lonny D. Garvey
Director of Publications
National Tooling and Machining Association

MESSAGE TO EDUCATORS AND TRAINERS

The National Tooling and Machining Association (NTMA) has always been an advocate of quality training at the highest level. NTMA entered the textbook publishing business in 1964 with a few basic industry related textbooks.

In 1984 NTMA introduced the Metalworking Training System (MTS). Over $300,000 were spent to put together this modulerized training program. The uniqueness of the MTS is its adaptability to virtually any metalworking training program. Also, the MTS is the only training system recognized by the industry and approved by the National Institute for Metalworking Skills (NIMS).

The establishment of NIMS was the direct result of NTMA's extensive work and dedication to the industry. NIMS has developed the testing required for certification.

I personally invite you to contact the NTMA Publications Department for an MTS sample kit. NTMA has training experts that will work with you to improve your training program without "re-inventing the wheel."

Lonny D. Garvey
Director of Publications

800-248-6862 ● 301-248-2755 (fax)
or
Publications only-800-832-7753

TABLE OF CONTENTS

Chapter I

INTRODUCTION TO SHOP PRINTS

As a machinist, toolmaker, or diemaker, you will be working constantly with engineering drawings and layout procedures. An engineering drawing is a collection of straight lines, curves, and figures that shows the shape and dimensions of an object in such a way that—if you read and interpret the drawing properly—you can make the object or part exactly as the designer intended. Metal parts, whether simple or complex, must be machined to exact shapes and dimensions. It would be practically impossible for a designer to describe the shapes and dimensions of an object either orally or in writing, so he depends on a very accurate drawing to convey his ideas to you.

Thus you must be able to read and interpret a drawing accurately, so that you can determine what you must do to make the metal part. Then, using various layout tools and measuring devices, you must be able to lay out the job on the metal blank. Finally, you must be able to perform the machining operations necessary to produce the finished object. This is the sequence in which nearly every machining job is done: drawing interpretation, layout work, and machining operations. In this chapter we will focus upon the kinds of drawings you will use in your work. In a later chapter, you will learn about layout tools and layout techniques.

WHAT IS A BLUEPRINT?

A blueprint is one of several ways that engineering drawings are reproduced. The engineering drawings made by either the design engineer or a draftsman go to the shop in the form of prints made from the original drawings. Several different printing processes are used to reproduce the drawings.

How a Blueprint Is Made

The term blueprint originally referred to a negative contact reproduction process in which the white areas of the original drawing become blue and opaque, and the dark areas or linework on the original drawing remain white. A blueprint is made by first placing a piece of transparent tracing paper or cloth over the original drawing, and making an exact tracing of the drawing. The tracing is then placed against a glass plate held in a framework, and a sheet of chemically-coated paper is placed over it. The back of the boxlike frame is closed, with pressure springs holding the glass, tracing cloth, and treated paper tightly together.

Then the frame is exposed to ultraviolet light. The light that passes through the transparent portions of the tracing causes the paper to turn a pale blue color. The portions not exposed to the light (the lines, letters, figures, etc., on the tracing) remain white. The treated paper is developed in a processing solution that intensifies and fixes the blue areas, and allows the unexposed areas to remain white. When the print has dried, it is ready to use. The entire procedure takes several minutes to complete. Because of its blue color with contrasting white lines, the reproductions are called *blueprints.*

Nowadays, modern technological innovations have rendered the original blueprinting process virtually obsolete by developing easier and faster techniques to produce reproductions used by the machinist. Although they are not technically blueprints, the term is still used when referring to almost any type of reproduced drawing used on the shop floor. Some of these new methods include:

Vandykes—a positive print with brown lines on white paper,

BW Prints and Directo Prints—black lines on white paper,

Ozalid Prints—black, blue, or dark red lines on white paper,

Photostats—white lines on a dark blue paper —when photostated a second time—brown-black lines on white paper,

Lithoprints—black lines on white paper, and

Photocopies—black lines on white paper.

Classification of Engineering Drawings

Engineering drawings that provide all the information necessary to manufacture an object or a part are called *working drawings.* A set of working drawings in the metalworking industry is usually divided into two general classes: assembly and detail. A set of working drawings includes:

(1) The full graphic representation of the shape of the object or part,

(2) The figured dimensions of the object or part,

(3) Explanatory notes (both general and specific) on the individual drawings giving the specifications of material, heat treatment, finish, etc.

(4) A descriptive title on each drawing,

(5) A description of the relationship of each part to the others in an assembly, and

(6) A parts list or bill of material for an assembly.

Assembly drawings are of two types: main assembly and subassembly. The *main assembly drawing* shows a finished object, indicating all parts in their proper position and in relation to one another. Only principal dimensions dealing with assembly or installation are indicated. A *subassembly drawing shows two or* more parts joined together in the shop assembling procedure to form a unit of an object or piece of equipment, which is not itself a complete assembly.

Detail drawings describe an individual part, and contain all the information necessary to manufacture that part.

Title Blocks on Working Drawings

Although the lines and dimensions on a drawing provide complete descriptions as to the size and shape of an object, they cannot furnish all of the information you need to make the part or object. The engineer or draftsman must provide this essential information without adding to the complexity of the drawing; yet it must be placed on the drawing where it can be located easily and interpreted correctly. Therefore, such information is usually placed in the title block, the revision box, or in notes placed so that they do not interfere with the drawing itself. By knowing what a title block consists of, how notes are used, and how revisions and changes are indicated, you will be able to obtain all the information you need from a drawing.

The title block, usually located in the lower right-hand corner of a drawing, contains information not directly related to the construction of an object, but which is necessary for its manufacture. The block is divided into sections which generally give the following information:

(1) Name of the company and its location

(2) Name of the part

(3) Part number, die number, forging number, etc.

(4) Drawing number assigned to the part number

(5) Number of parts required for each assembly

(6) Scale indicating the size of the drawing compared with the actual size of the part. Detail drawings are usually made full size. Large

parts and assemblies may be drawn to a reduced scale to fit on the paper. Very small parts may be drawn two or three times their actual size to show details clearly. The most common scales are full (actual) size, and 2, 4, ½, and ¼ times actual size.

(7) Assembly drawing number (on a detail drawing) to identify the part in the assembly

(8) Drafting room record, which includes the names or initials (together with date signed) of the persons responsible for the accuracy of the drawing: draftsman, checker, engineer, examiner, and production approval authority

(9) Material or materials to be used in making the part, together with optional materials. Reference is usually made to notes because of the limited space in the block.

(10) Stock form and/or size

(11) Heat treatment required during the manufacturing process (if any)

(12) Protective treatment code specifications

(13) Tolerances (general) that apply to all dimensions that do not have individual tolerances included with the basic dimensions.

(14) Finish marks indicating which surfaces must be machine finished. A modified check mark (√) containing a digit or digits is the controlled symbol used to indicate the degree of finish.

(15) Shop notes (general and specific) provide information and instruction that cannot be given conveniently by any other means. *General* notes apply to the drawing as a whole. *Specific* notes apply to particular parts of the drawing, and are located near the area of the drawing to which they pertain.

(16) Drawing revisions and/or changes call attention to variations in the original design caused by unsatisfactory performance or difficulty in manufacturing. These changes are usually located in the upper right-hand corner of a drawing in a separate box called the *Revision Box.* All changes to the drawing are entered, dated, and identified by a number or a letter. (Sometimes a revised drawing is made instead, and a letter indicating that it is a revised drawing is added to the original drawing number.)

THEORY OF PROJECTION OF DRAWINGS

In blueprint reading, a view of an object is known technically as a projection. A *projection* is an extension of lines of sight (called *projectors*) from the eye of the observer through lines and points on the object to *the plane of projection.* The two main classes of projection are central or perspective projection, and parallel projection.

Central Projection

In central or perspective projection, the projectors converge to a point, as shown in Fig. 1-1, which is the eye of the observer. Therefore, each projector forms a different angle with the plane of projection. The result is that the size of the view varies, depending on how far away the observer is from the object, and how far the object is from the plane of projection. Because central projection distorts the actual shape of an object or part, and

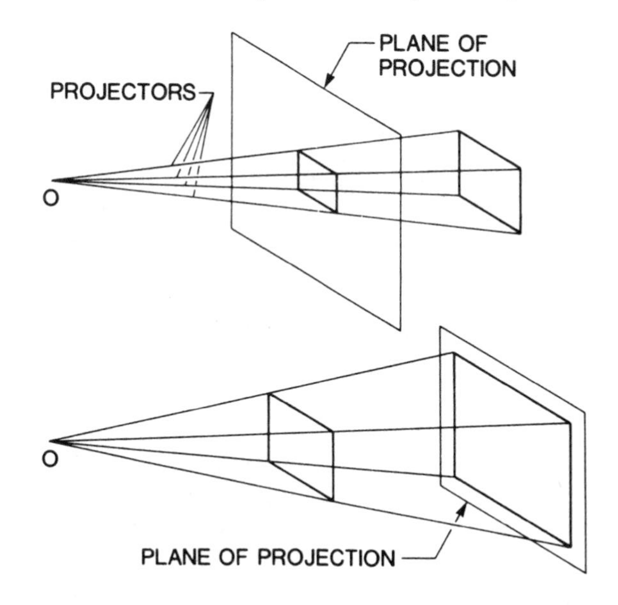

Figure 1-1 Central projection.

does not show its true shape, it is not used on engineering drawings, except in preparing sketches. Parallel projection is the usual practice on engineering drawings.

Parallel Projection

In parallel projection, instead of converging to a point, the projectors remain parallel to each other, as shown in Fig. 1-2, giving the true size and shape of an object. As shown in Fig. 1-2, when the projectors are perpendicular to the plane of projection, the size and shape of the object on the flat plane of the drawing paper are represented as they actually are, regardless of relative positions and distances.

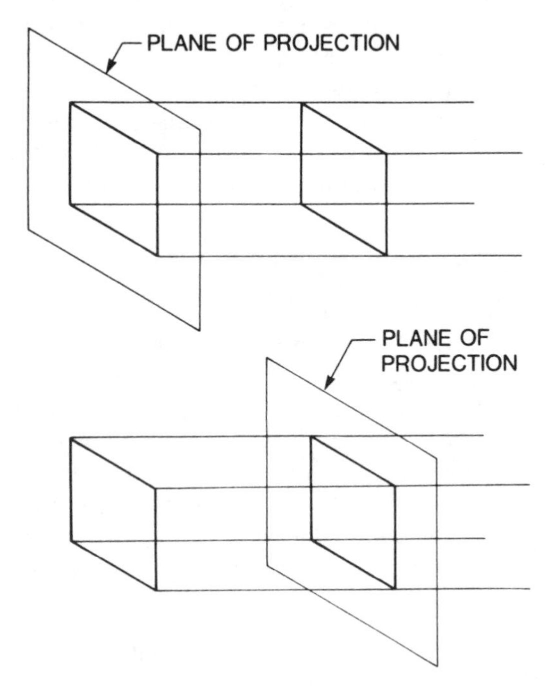

Figure 1-2 Parallel projection.

The two types of parallel projection used in engineering drawings are orthographic and pictorial or oblique.

Orthographic projection is a parallel projection in which the projectors are *perpendicular* to the plane of projection (see Fig. 1-2). Orthographic projection shows two dimensions—length and width.

Pictorial or oblique projection is a parallel projection in which the projectors are slanted or *at an angle* to the plane of projection. Pictorial projection shows three dimensions—length, height, and width. It is one way to show all three dimensions in a single view.

TYPES OF WORKING DRAWINGS

The two types of working drawings that you will use are called *orthographic projections* and *pictorial drawings*. The relationship between them is shown in Fig. 1-3. View *A* is a pictorial drawing of a rectangular box. The two vertical planes of the box in view *B* are revolved about their respective lines of intersection with the horizontal plane, so that all three views (top, front, and end) are shown in a single plane—the plane of the paper as in view *C*. View *C* is called an orthographic projection.

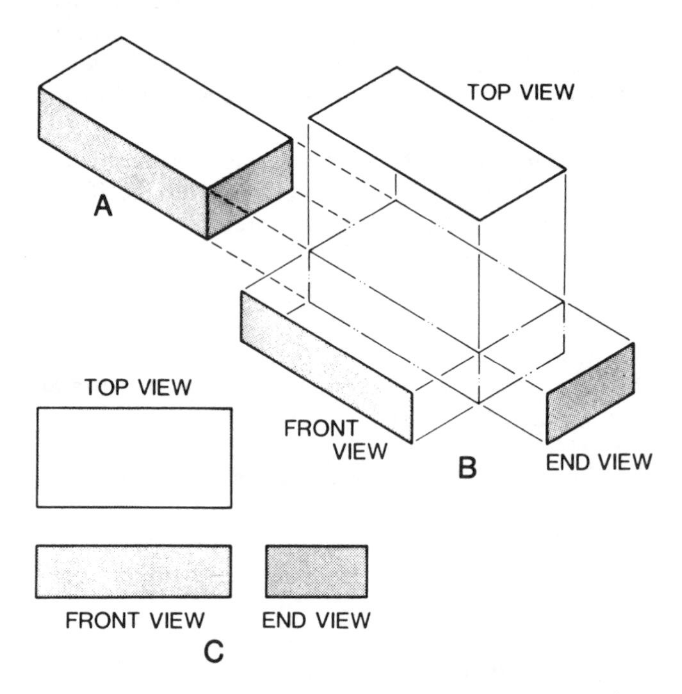

Figure 1-3 Pictorial and orthographic projection drawings.

In the pictorial view (*A*), it appears that the angles at the four corners are not equal, whereas in the orthographic projection, we can see that all of the angles are equal. Pictorial drawings always show the features of an object somewhat distorted, and are used only to a limited extent in working drawings. Orthographic projections, on the other hand, show all of the features of an object in their true shape, and are almost always used to make detail drawings. Three views of the object are usually given to show the three main dimensions of the object: length or height, width, and depth or thickness.

Orthographic Projections

The orthographic projection showing the top is called the *top view.* It shows that part that you would see if you were standing directly over the object, and looking down at it.

The orthographic projection showing the front is called the *front view.* It shows that part of the object that you would see if you were standing directly in front of the object, and looking at it squarely.

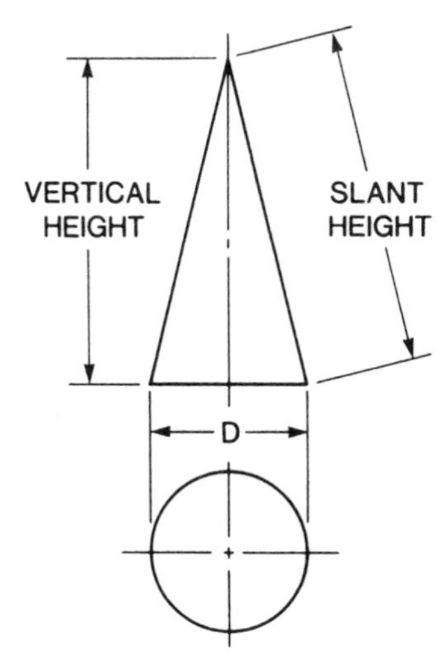

Figure 1-4 Orthographic projection of a cone.

The orthographic projection of the *end view* shows the distance from the front to the hidden back view. It also shows the part of the object that you would see if you were standing at the end and looking at the object squarely. In simple orthographic projections of regular shapes like the rectangular box in Fig. 1-3, the end view is not important and may be omitted, because all dimensions can be obtained from the top and front views.

Curved surfaces do not always look curved in an orthographic projection, because you are viewing only the top, bottom, end, or side of an object 90 deg to the surface. In a pictorial drawing, you can see that the edges are curved, but in an orthographic projection, a curved surface appears flat. Figure 1-4 is a two-view orthographic projection of a cone. Although you know that the surface of a cone is curved, you cannot see the curvature in the drawing. However, from the adjacent end view, you can see that the base is circular and, therefore, that the surface of the cone is curved. Keep in mind that orthographic projections do not show curved surfaces, but will indicate a curved surface in one of the views.

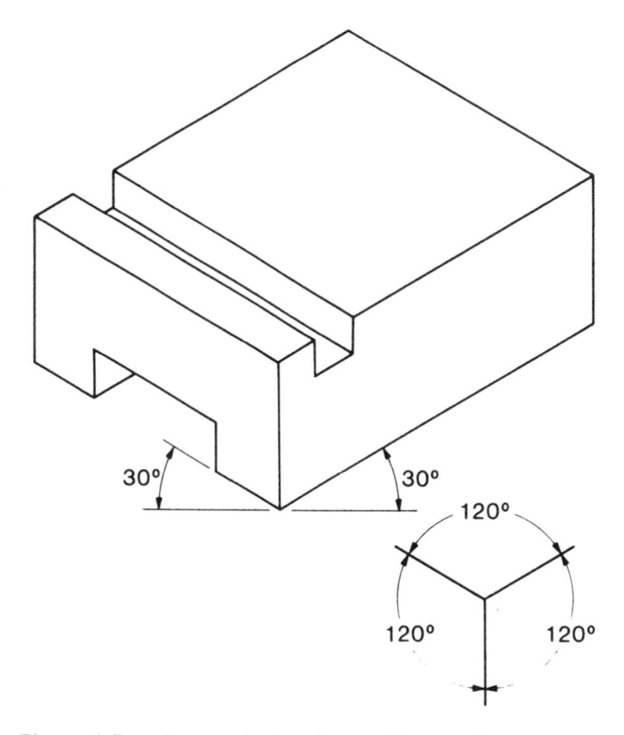

Figure 1-5 Isometric drawing and isometric axes.

Pictorial Drawings

The basic purpose of a pictorial drawing is to describe an object by showing all three of its dimensions in one view, instead of in three separate views as in orthographic projections. Pictorial drawings are sometimes included in working drawings to show the general location, function, and appearance of parts and/or assemblies. The two main types of pictorial drawings used in engineering drawings are isometric and oblique.

Isometric drawings represent an object in three dimensions, drawn about the three isometric axes that are 120 deg apart (see Fig. 1-5). All lines that are parallel on the object are also parallel on the drawing. Vertical lines are shown vertically, but horizontal lines are drawn at an angle of 30 deg to the horizontal. All the lines that represent the horizontal and vertical lines on an object have true length. Isometric drawings may be dimensioned, and blueprints of these drawings can be used to make simple objects or parts. However, isometric drawings alone do not provide enough information to make complex parts or structures. They are sometimes used in conjunction with orthographic projections to clarify and aid interpretation.

Oblique drawings also show an object in three dimensions (see Fig. 1-6). However, in an oblique

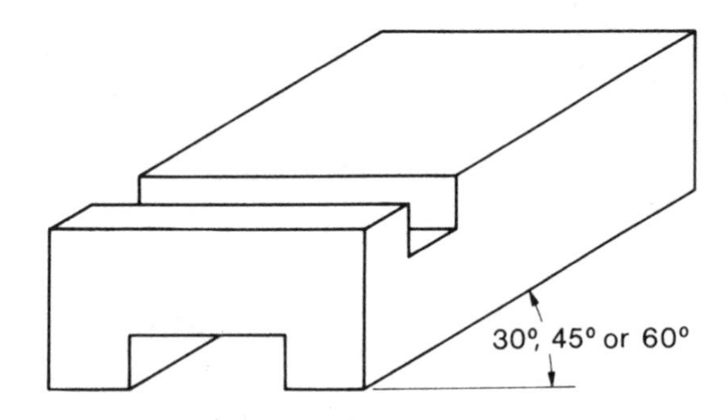

Figure 1-6 Oblique drawing.

drawing, only the front of the object is shown in its true size and shape, as in an orthographic projection. But the receding lines of the other two sides are drawn obliquely at any angle—usually 30, 45, or 60 deg to the horizontal. Although the front view is shown in its true size and shape, the receding sides are scaled at less than their actual length—normally ¾ the scale of the front view.

Cabinet drawings are a special type of oblique drawing that shows the sides of an object at one-half their scale length, using 30 or 45 deg angles.

Cavalier drawings are another special type of oblique drawing that shows the oblique sides drawn to the same scale as that used on the front view. This creates a distortion from the object's true proportions, but allows the use of one scale of measure for the entire drawing. The oblique sides are drawn 45 deg to the horizontal.

Compare the differences between cavalier, cabinet, oblique, and isometric drawings shown in Fig. 1-7.

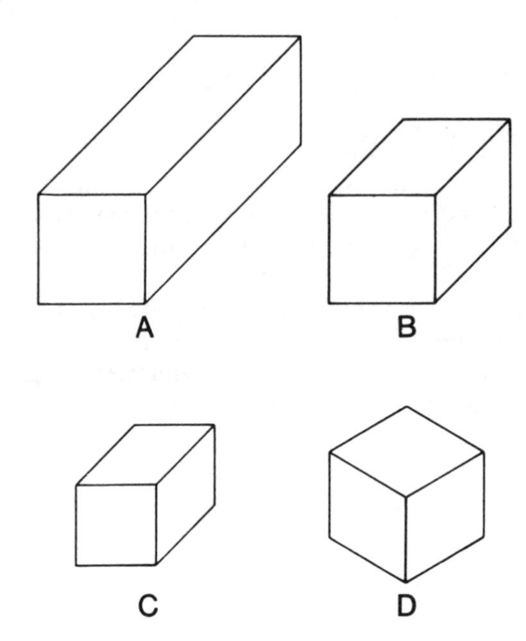

Figure 1-7 Pictorial drawings: (A) cavalier, (B) cabinet, (C) oblique, (D) isometric.

SUMMARY

A blueprint is a reproduction of an engineering drawing, which is a graphic form of communication from the engineer/designer to you telling you exactly what an object looks like. Blueprints are made in a three-step process: (1) drawing, (2) ink/pencil tracing, and (3) the print, a photographic reproduction of the tracing.

Besides the lines, curves, and figures in an engineering drawing, additional information is contained in the title block.

The two classifications of working drawings that you will use are assembly drawings and detailed drawings. Machinists use detailed drawings most often, because of the amount of flexibility inherently built into them, and the vast amounts of detail that can be drawn on the print with relative clarity.

TRAINING PRACTICE

1. Define the term *engineering drawing* in your own words.

2. Define the term *blueprint* in your own words.

3. List the three steps in making a blueprint.

4. Where on a drawing would you find the part name, material used, part number, etc.?

5. Where on a drawing would you look for notes on recent changes, and what is the box called?

6. What is the correct name of a drawing that has only one view showing length, width, and depth?

7. Which type of print shows all of the components of a mechanism put together in proper location?

8. A three-view drawing of a single part is called a/an

Chapter II

LINES AND THEIR USES IN ORTHOGRAPHIC PROJECTIONS

Your ability to read and interpret drawings and blueprints depends on your ability to recognize the different types of lines used in making the drawings, and to understand how they serve to describe the object or parts represented. Lines used to represent an object and to aid in reading the drawing are made in definite standard forms as illustrated in Fig. 2-1.

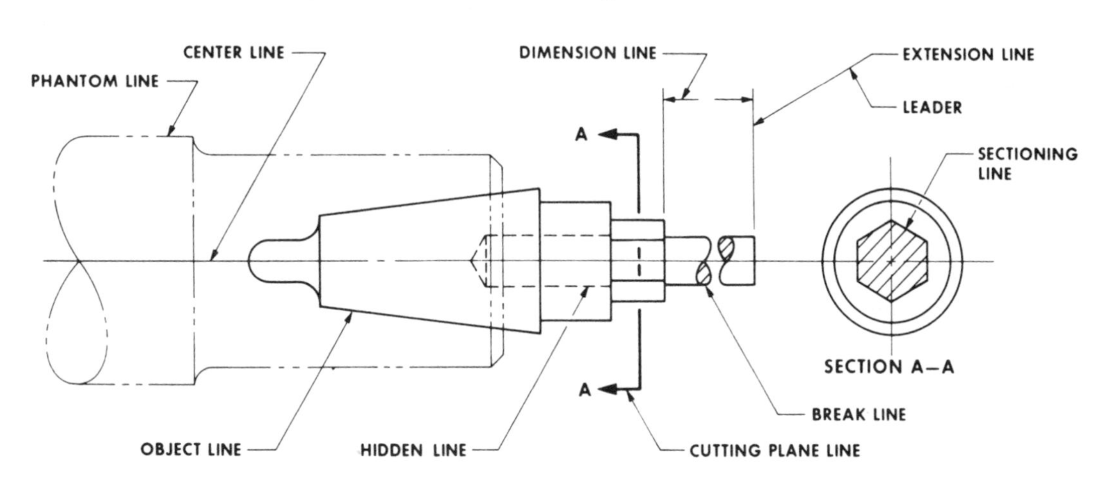

Figure 2-1 Family of lines.

FAMILY OF LINES

The relative thickness of a line (thick or thin) and its composition (solid, broken, dashed, etc.) have specific meanings. Because of their fundamental importance to blueprint reading, these lines are called the *family of lines.*

Object Lines

Object or visible lines are thick, solid lines that outline all surfaces visible to the eye. These are the most important in the family of lines. Being thick and solid, they become the basis for comparing the weights and composition of all other lines.

THICK and SOLID

Figure 2-1A Object lines.

Hidden Lines

Hidden or invisible lines, consisting of short evenly-spaced dashes, outline invisible or hidden surfaces. They are thin lines, about half as heavy as visible lines. They always begin with a dash in contact with the line from which they start, except when a dash would form a continuation of a solid line.

THIN and DASHED

Figure 2-1B Hidden or invisible lines.

Centerlines

Centerlines consist of alternating long and short evenly-spaced dashes, with a long dash at each end and short dashes at points of intersection. The lines are the same weight as invisible lines. Centerlines indicate the central axis of an object or parts, particularly circular objects or objects made up of circular or curved parts. They are also used to indicate the travel of a center. Whenever a complete circle or hole is shown on a drawing, both horizontal and vertical centerlines are used to indicate the center point of the circle or hole. Centerline is often abbreviated C/L or the symbol ℄ is used.

THIN and ALTERNATING LONG and SHORT

Figure 2-1C Centerlines.

Phantom Lines

Phantom lines are thin lines used to indicate alternate positions of the parts of an object, repeated detail, or the locations of absent parts. They are made by alternating one long and two evenly-spaced short dashes, with a long dash at each end.

THIN and ALTERNATING ONE LONG and TWO SHORT

Figure 2-1D Phantom lines.

Dimension Lines

Dimension lines are short, solid lines that indicate the distance between two points on a drawing. They terminate in arrowheads at each end, and are broken to insert the dimension.

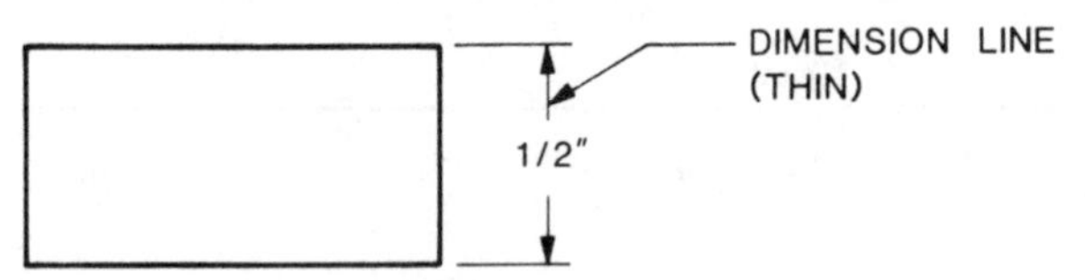

Figure 2-1E Dimension lines.

Extension Lines

Extension lines are short, solid lines used to show the limits of dimensions. They may be placed inside or outside the outline of an object. They extend from an outline or surface, but do not touch it. Extension lines are the same weight as invisible lines.

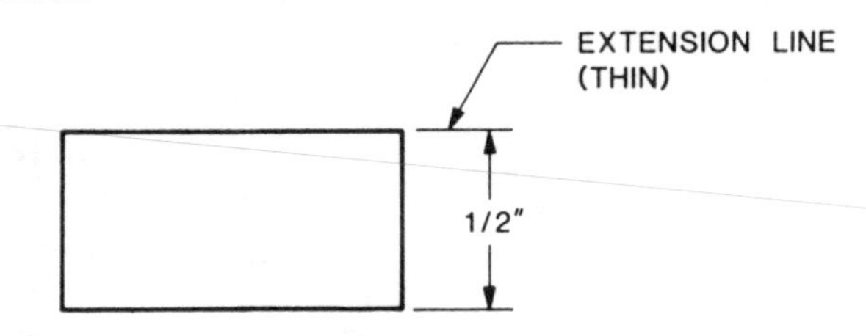

Figure 2-1F Extension lines.

Leaders

Leaders or leader lines indicate the part or area of a drawing to which a number, note, or other reference applies. They are thin, solid lines and usually terminate in a single arrowhead.

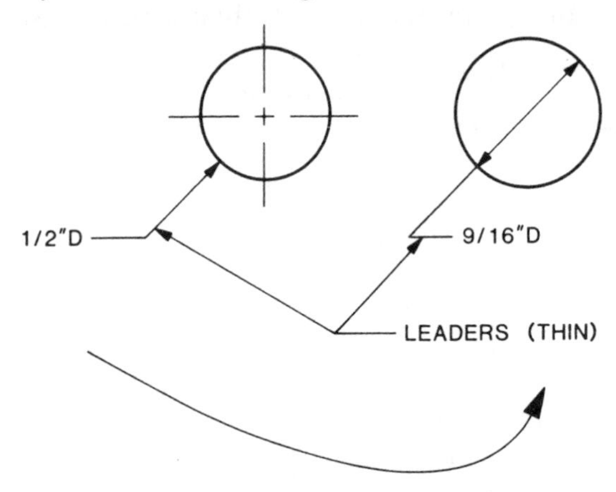

Figure 2-1G Leaders.

Break Lines

Break lines (both long ones and short ones) indicate that a part is broken out or removed, either to (1) show more clearly the part or parts that lie directly below the broken out part, or (2) to reduce the size of the drawing of a long part having uniform cross section so that it can be shown on a smaller sheet of paper. Short breaks are indicated by solid, thick, freehand lines. Long breaks are indicated by solid, thin, ruled lines broken by freehand zigzags. Breaks on shafts, rods, tubes and pipes are curved (see Fig. 2-1H(a)), and are made with a draftsman's tool called a *french curve.* Breaks on rectangular parts or part details are made freehand, as shown in Figs. 2-1H(b) and (c).

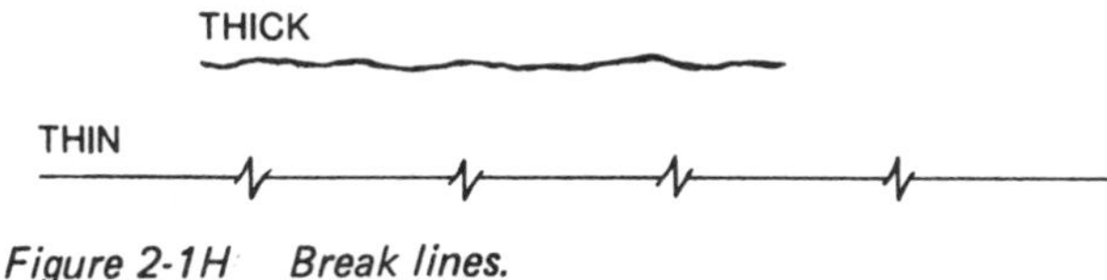

Figure 2-1H Break lines.

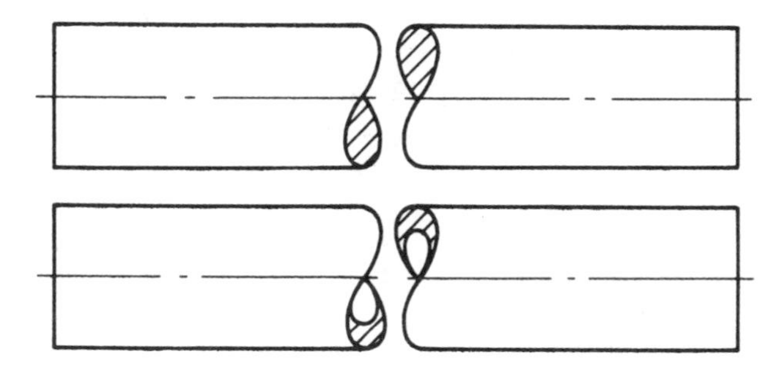

Figure 2-1H(a) Round solid, hollow cross sections.

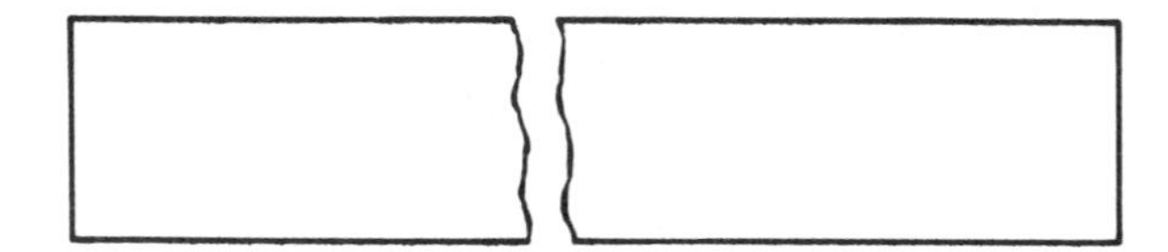

Figure 2-1H(b) Rectangular parts.

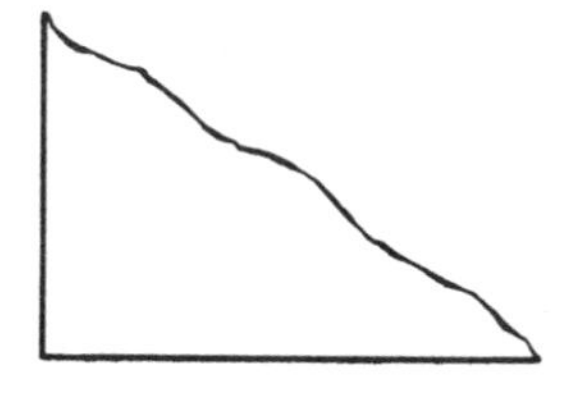

Figure 2-1H(c) Part details

Section Lines

Section lines or crosshatch lines distinguish between two separate parts that meet at a given point. Each part is lined or hatched in opposite directions with thin parallel lines placed approximately 1/16 in. apart at 30 deg, 45 deg, or 60 deg across the exposed cut surface. Most section lines are drawn as shown in Fig. 2-1I.

However, sometimes section lines are used to depict a particular type of material. Although the section lines in Fig. 2-1I represent cast iron, they are used generally to show any cut or sectioned surface. In assembly drawings, where many kinds of materials may be used, individual parts may be crosshatched with the symbol for a particular material. Some of the commonly-used material symbols are shown in Fig. 2-1J.

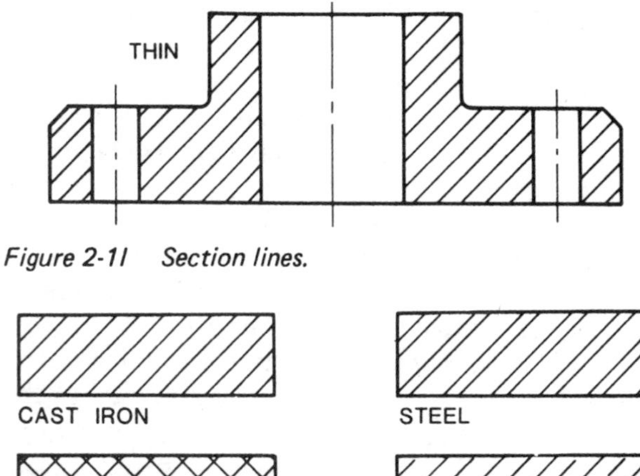

Figure 2-1I Section lines.

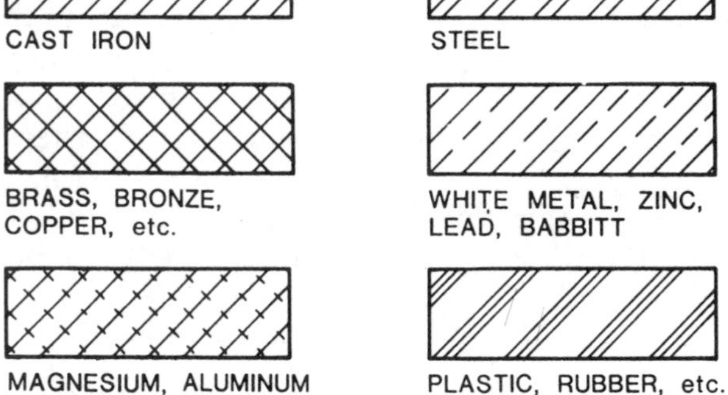

Figure 2-1J Symbols for common materials.

Cutting Plane Lines

A cutting plane line consists of a heavy dash followed by two shorter dashes. At each end, it has a short line at right angles to the cutting plane line terminating with arrowheads pointing in the direction from which the cut surface is viewed. Cutting plane lines are usually labeled with a letter at either end to identify the drawing of the cut surface indicated by the same letters on the same sheet of paper. The cut surface drawing is called a *section.*

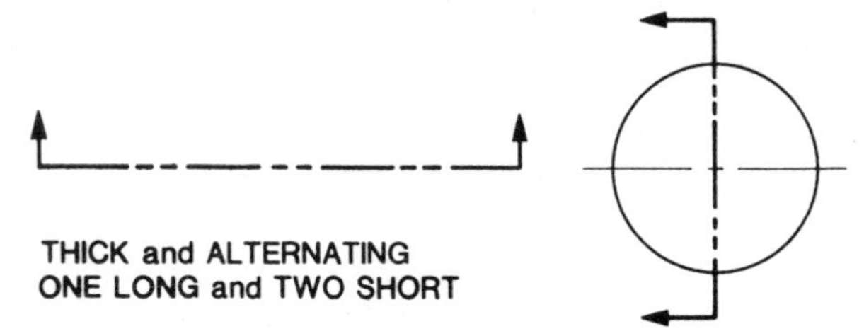

Figure 2-1K Cutting plane lines.

Screw Threads

Screw threads are shown on orthographic drawings using one of two methods: semi-conventional or conventional. Both replace the curved lines of the screw thread helix with straight lines. *Semi-conventional representation* (see Fig. 2-2) is used for thread diameters of 1 in. or more; *conventional representation* (see Fig. 2-3) is used for threads of 1 in. or less in diameter. The conventional representation approved by ANSI includes two series of thread symbols: regular (also called *schematic*) and simplified. The same symbols are used for all threads, regardless of their form.

In the *regular series* (see Fig. 2-3A), long vertical lines indicate crests, whereas short vertical lines indicate roots. The thread profile itself is usually not shown. In the *simplified series* (see Fig. 2-3B), the threaded portion is indicated by dashed lines drawn parallel to the axis at about the depth of the thread. The detailed specifications for the screw thread are given in a specific note tied by leaders to the drawing of the fastener itself. Screw thread specifications are written, in sequence: nominal size, number of threads per inch, thread series symbol, thread class symbol, and hand.

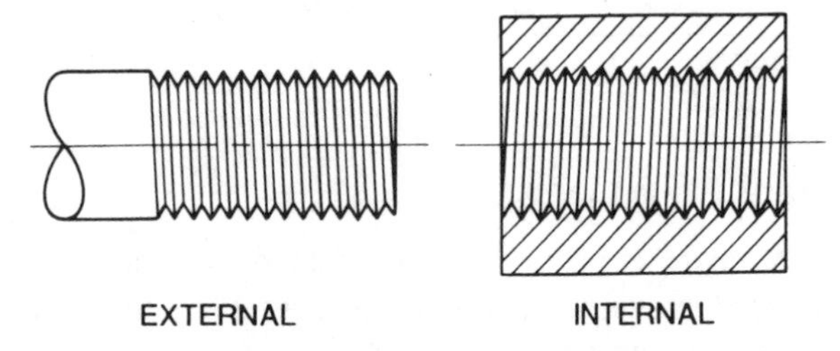

Figure 2-2 Semi-conventional representation of screw threads.

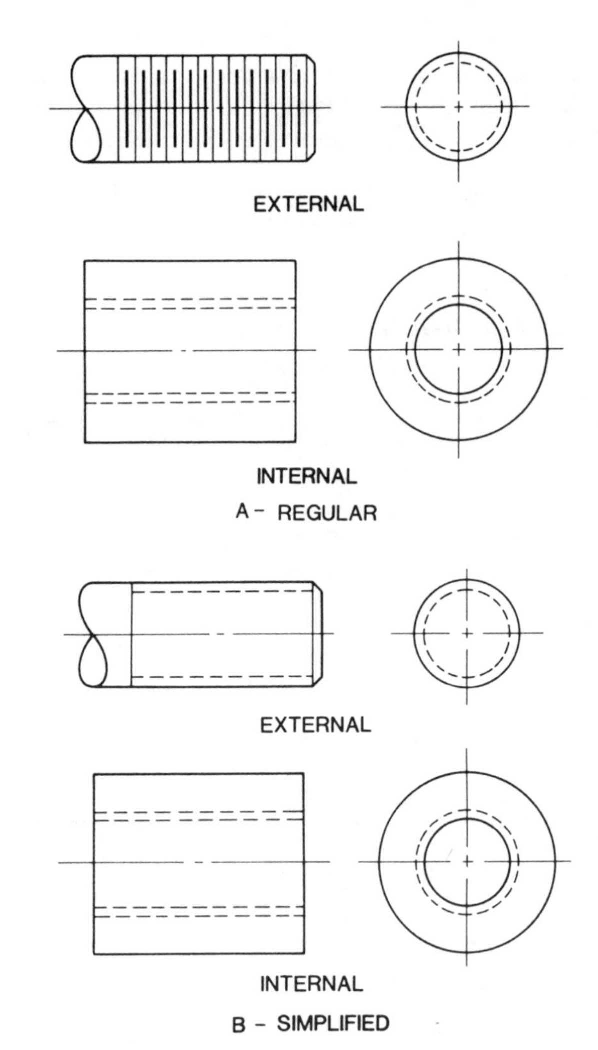

Figure 2-3 Conventional representation of screw threads approved by ANSI.

PRINCIPLES OF ORTHOGRAPHIC PROJECTION

You learned in Chapter I that the primary system of engineering drawing used in the United States is known as orthographic projection. Recall that an *orthographic projection* is a view of an object drawn on a plane, known as the *plane of projection. Projectors* are the parallel lines of sight that connect all corners and edges of the object to the plane of projection. In orthographic projection, the projectors are perpendicular to the plane of projection, or at right angles to it.

Multi-View Projection

A single orthographic projection of one face of a cube could not describe the shape of that object completely. Therefore, more than one projection and more than one plane of projection must be shown on a drawing to indicate that it is a cube. This is called *multi-view projection* and is the simplest way to represent objects.

In multi-view projection, the three principal planes of projection are situated at right angles to each other, and are called the *vertical, horizontal,* and *profile planes* as shown in Fig. 2-4. They intersect in straight lines called the *coordinate axes.* The four angles formed by the intersection of the vertical and horizontal planes are called, as

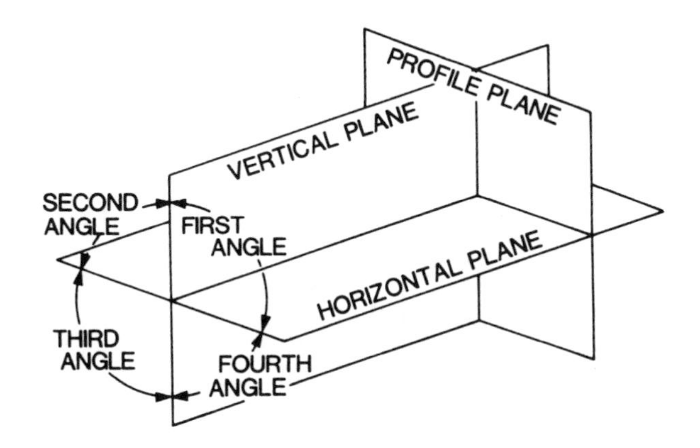

Figure 2-4 Multi-view planes of projection.

shown in Fig. 2-4, *first, second, third* and *fourth angles.* These angles are considered to revolve clockwise around one axis, much as the Earth revolves around its own axis. The object to be projected can be located in any one of the four angles. However, for technical reasons, the second-angle and fourth-angle projections are almost never used in engineering drawings.

Third-Angle Projection

In the United States and Canada, third-angle projections (see Fig. 2-5) are the only ones used on drawings and blueprints produced in those countries. As you can see, they project a front view on the vertical plane, a top view on the horizontal plane, and a right side view on the profile plane—each surface shown in its true shape as it would appear to an observer looking directly at it.

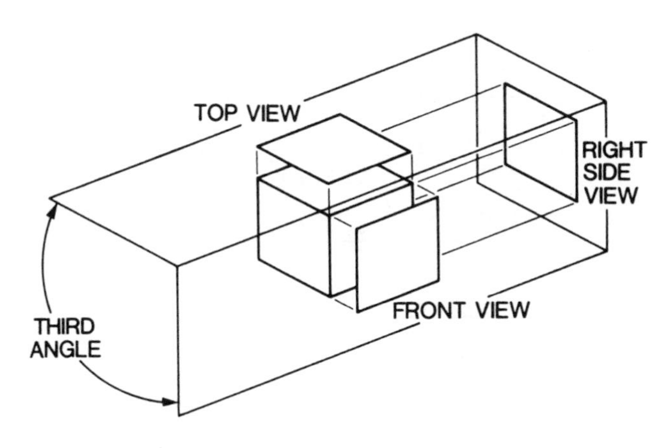

Figure 2-5 Third-angle projection of a cube.

First-Angle Projection

First-angle projection is used in European drafting practice. You will learn more about it in Chapter VI.

View Alignment

Any cube or rectangular object can project six views, because the object has six sides or surfaces: front, back, top, bottom, left side, and right side as shown in Fig. 2-6A. If you took the seams of the cube apart so that it lay flat, each side would be projected or aligned at right angles, as shown in Fig. 2-6B.

To further illustrate view alignment (or the standard arrangement of views on a drawing),

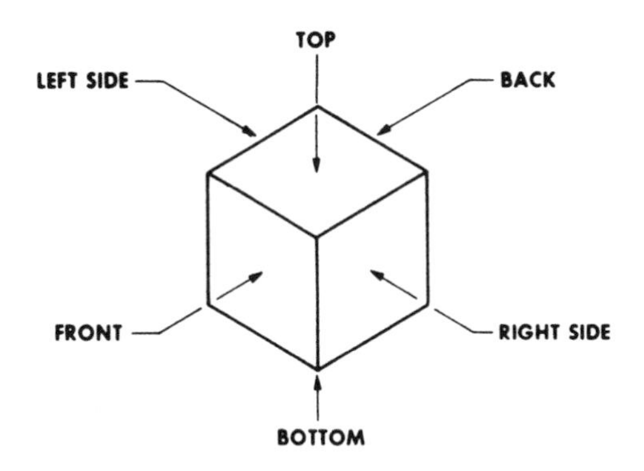

Figure 2-6A Cubes project six views.

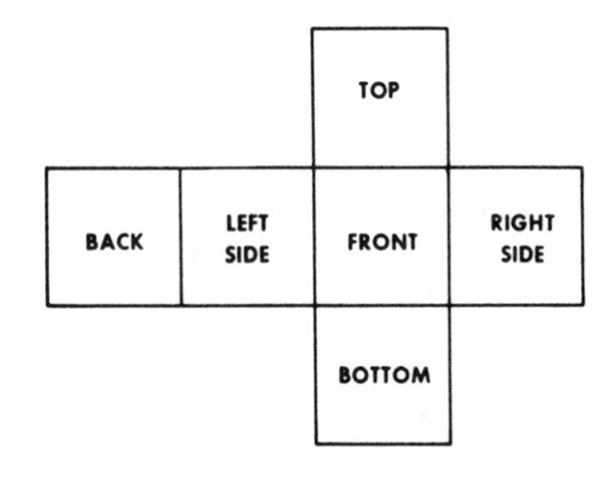

Figure 2-6B Sides of cube laid flat.

Fig. 2-7A shows a third-angle projection of a part inside a glass box with hinged seams. Each view of the object is projected on the corresponding face of the box. Now, imagine rotating the hinged sides of the plastic box around clockwise, and flattening them out into a single plane, such as a piece of drawing paper. Note in Fig. 2-7B that:

(1) The top view aligns directly over the front view,

(2) The right side view appears directly to the right of—and in line with—the front view,

(3) The left side view appears directly to the left of—and in line with—the front view,

(4) The bottom view aligns vertically with the front and top views, and

(5) The back view appears to the left of—and in line with—the right side, front, and left side views.

Figures 2-6 and 2-7 show the standard alignment of the six possible orthographic projections, and although some views may be omitted if they are not necessary, no view should appear in any other position on a drawing. Note the dashed lines on the projections of the left side and bottom views in Fig. 2-7. It is customary to show features that are *behind* the front surface and *on* the bottom surface by dashed lines, indicating that they are hidden. However, if a feature that is hidden lies directly behind a visible feature, the two are represented by a solid line.

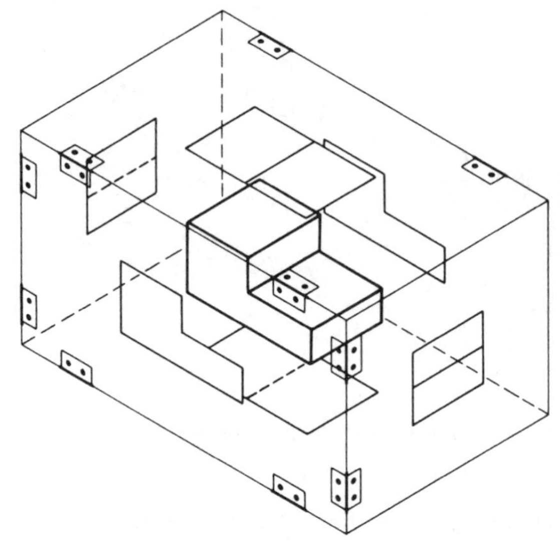

Figure 2-7A Third-angle projection of part in glass box.

In summary, remember that side views are always placed laterally (to the sides) of the front view in logical sequence. Top views are always above the front, and bottom views are always below the front.

Principal Views

The number of views presented on an ortho-

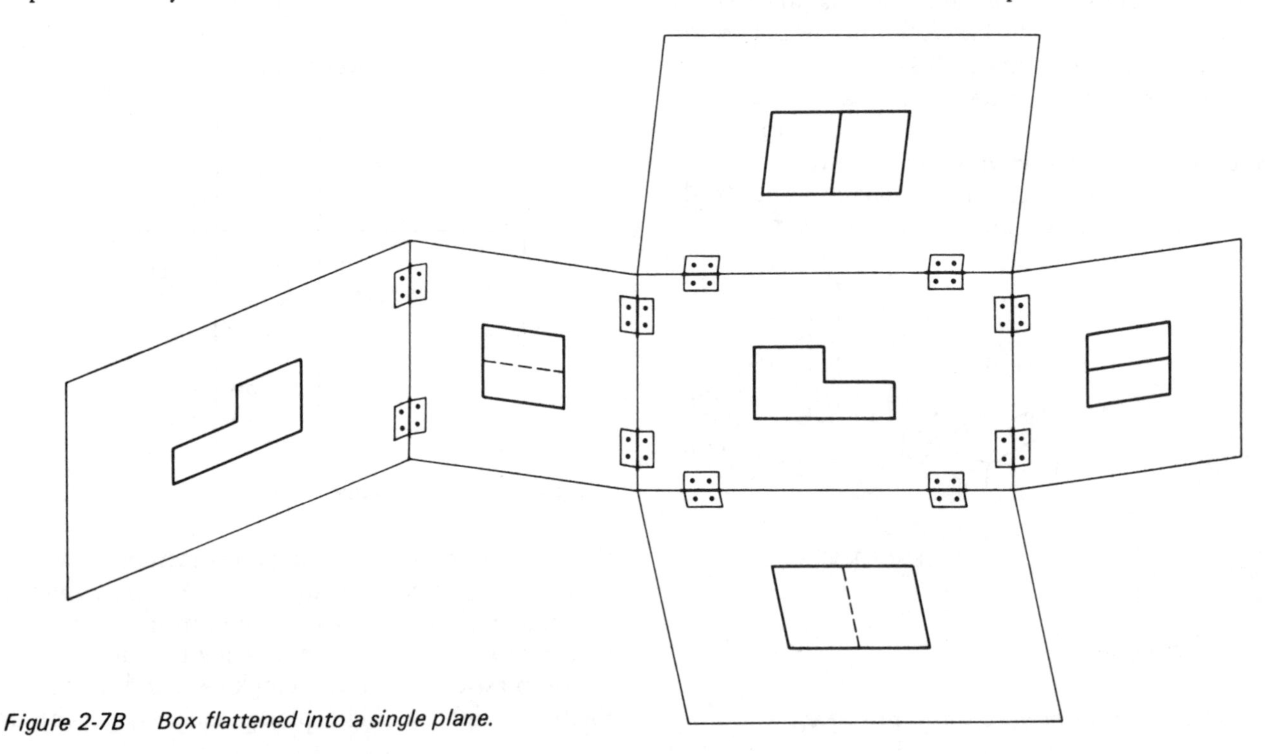

Figure 2-7B Box flattened into a single plane.

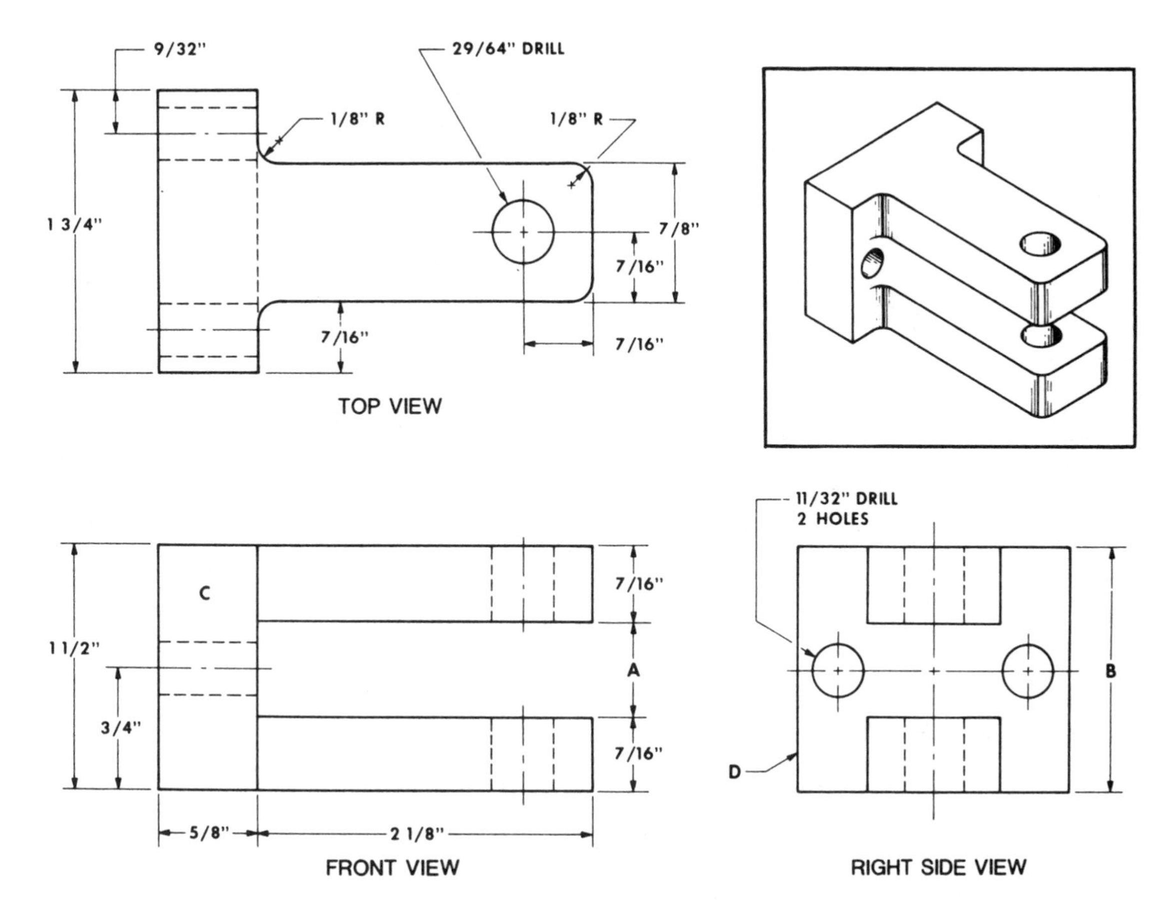

Figure 2-8 Views and their alignment on a detail drawing.

graphic drawing and their alignment help you in reading the drawing. Although detail drawings usually present only the three principal views of Front, Top, and Right side (see Fig. 2-8), the number of views necessary to make a part depends on the nature and shape of the object.

For example, from whatever angle you view a sphere, its shape is still the same. Therefore, unlike a cube or rectangular object, one view is sufficient to show the size, and the spherical shape can be explained in a short note.

Cylinders are often represented by one view also. Only a centerline running through the middle of the piece and the letter D (for diameter) in the dimensions are required to indicate that the piece is cylindrical in shape. However, any extra machining operations to be done on the cylinder, such as drilled holes, keyways, threads, and counterbores, would require a second—or even a third—view to accurately describe their size and location.

If an object has a complex or irregular shape, up to six views may be shown, and perhaps auxiliary views or sections. Auxiliary Views and Sectional Views are the subject of Chapter III.

Developing Orthographic Views

To make an orthographic three-view drawing from a third-angle projection, such as the pictorial

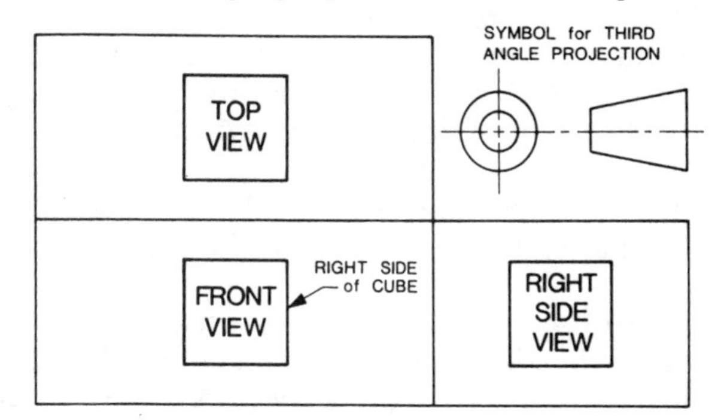

Figure 2-9 Third-angle projection of a cube brought into a single plane.

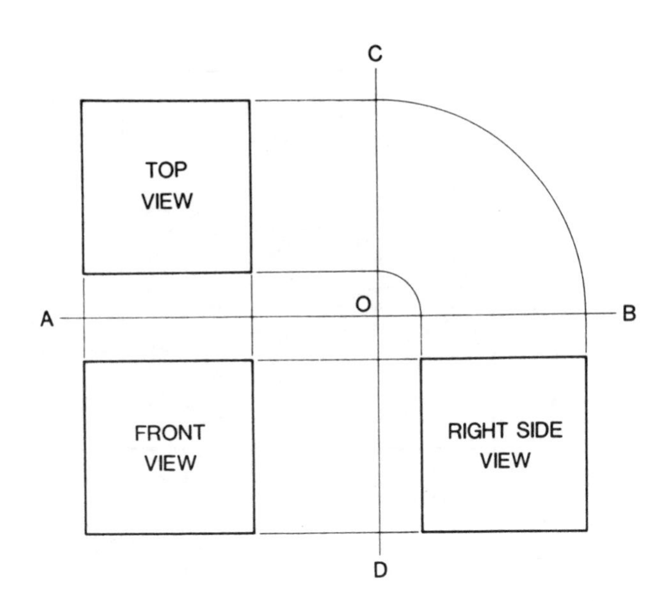

Figure 2-10 How to make a third-angle projection of a cube.

drawing of the cube in Fig. 2-6A, you would first assume that the vertical plane (front view) is the plane of the drawing paper. Then you would rotate the top and right side views clockwise. The projection would appear as shown in Fig. 2-9, which also shows the symbol for third-angle projection.

Figure 2-10 shows the basic principles of how to actually draw the projection shown in Fig. 2-9. First, draw horizontal Line *AB* and vertical Line *CD,* which intersect at *O.* Line *AB* represents the joint between the horizontal and vertical planes; Line *CD* represents the joint between these two and the profile plane. *Any two of the three views can be projected from whichever view is drawn first.* Assume, however, that you draw the front view first, and then project it upward, using vertical projectors, to draw the top view. Now project the top view to Line *CD* using horizontal projectors. With *O* as a centerpoint, use a compass to extend these projectors to Line *AB.* Draw the right side view by extending the projectors downward vertically from Line *AB,* and by projecting the right side of the front view to the right horizontally.

Two other ways to extend the top view projectors to the right side view are illustrated in Fig. 2-11. In Fig. 2-11A, the projectors are extended from Line *CD* to Line *AB* by lines drawn at 45 deg

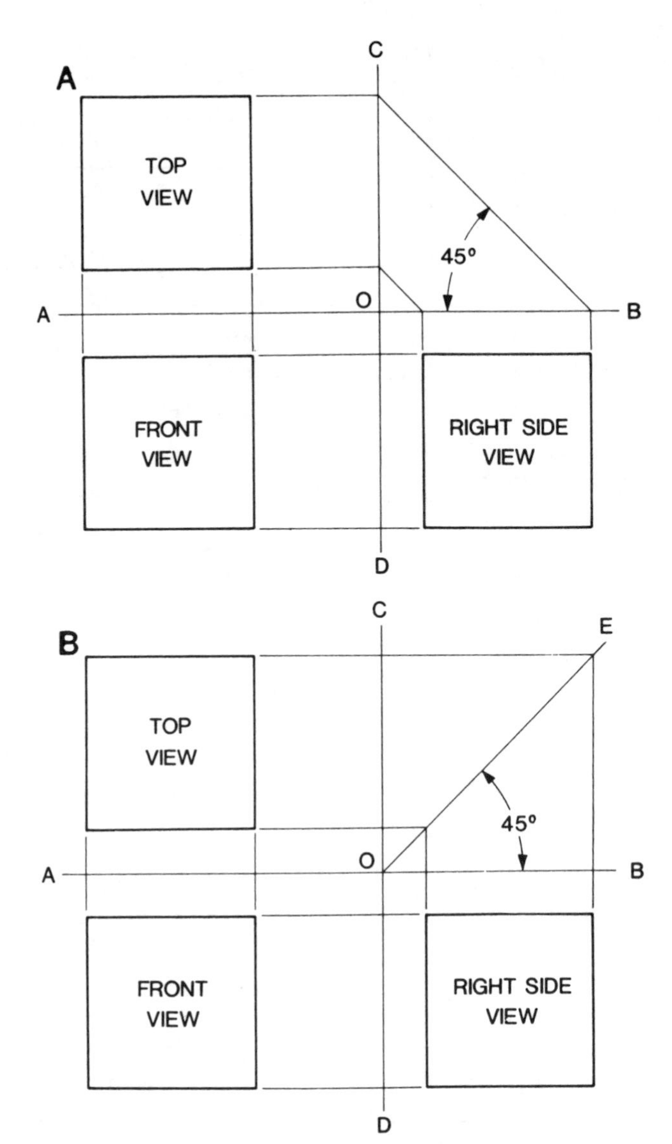

2-11 Alternate methods of extending top view.

to Line *AB.* In Fig. 2-11B, Line *OE* is drawn 45 deg to Line *AB,* and the top view projectors are first extended horizontally to Line *OE,* and then downward vertically.

Blueprint reading depends on your ability to visualize what an object will look like when rotated 90 deg horizontally and vertically. The ability to visualize the three principal views is a skill that you must develop. The techniques described above will help you with the visualization process, and also to construct third views from two given views using projectors.

SUMMARY

All orthographic drawings consist of a series of standard lines referred to as the family of lines. Each has a name, each is drawn differently, and each represents a different property on the blueprint.

Orthographic projection is right angle projection of an object. Six views are possible; usually only two or three are needed to describe an object.

You can find missing lines or make complete third views by projecting lines from one or two given views. The ability to construct and interpret views depends largely on your ability to visualize the object in its three-dimensional form. This is a skill you can develop by trying to create mental pictures of the part involved.

TRAINING PRACTICE

Part A (Missing Lines)

Complete the three-view drawings in the following exercises by adding any missing lines. The first one is done for you. Answers are found in Appendix A.

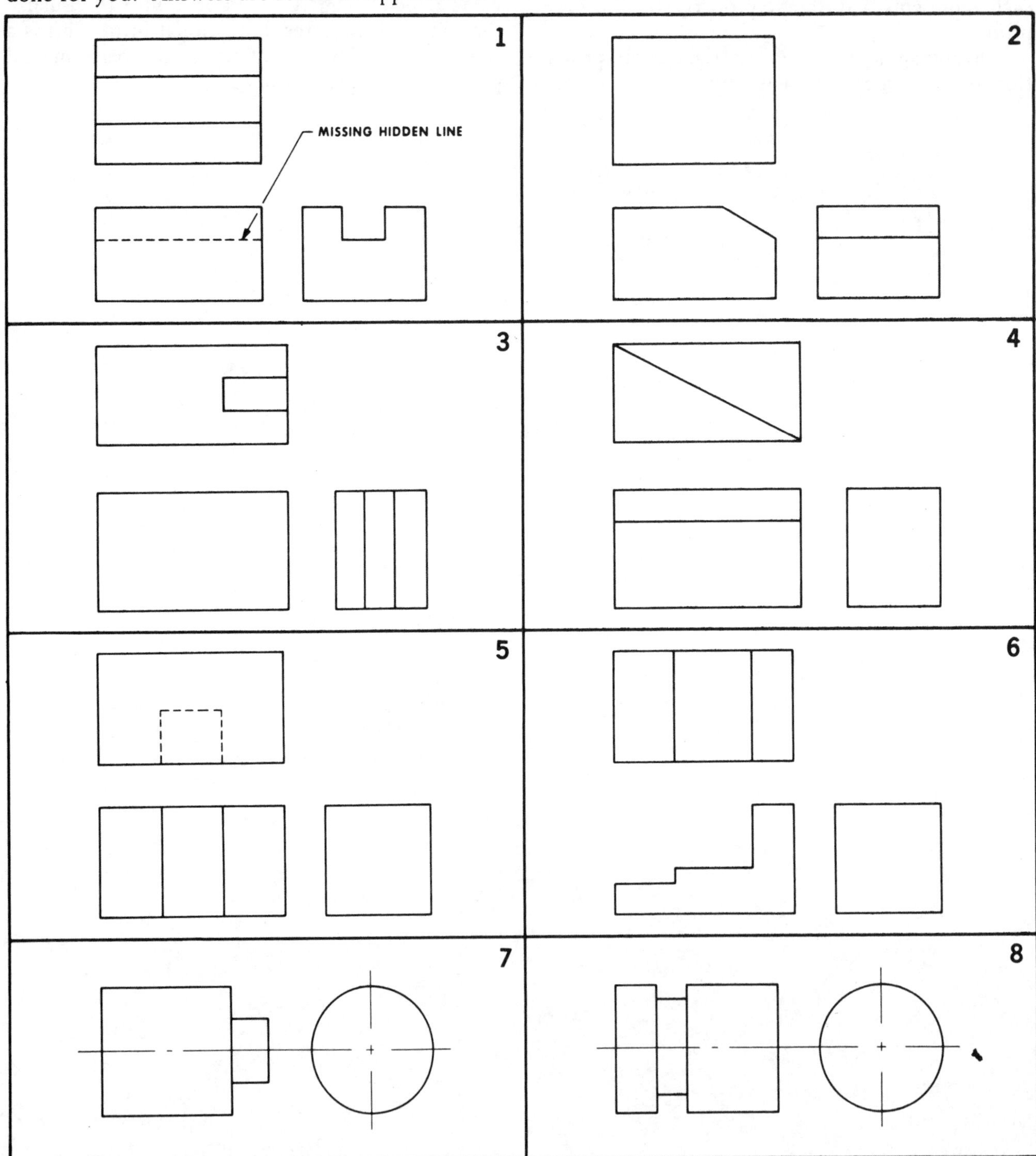

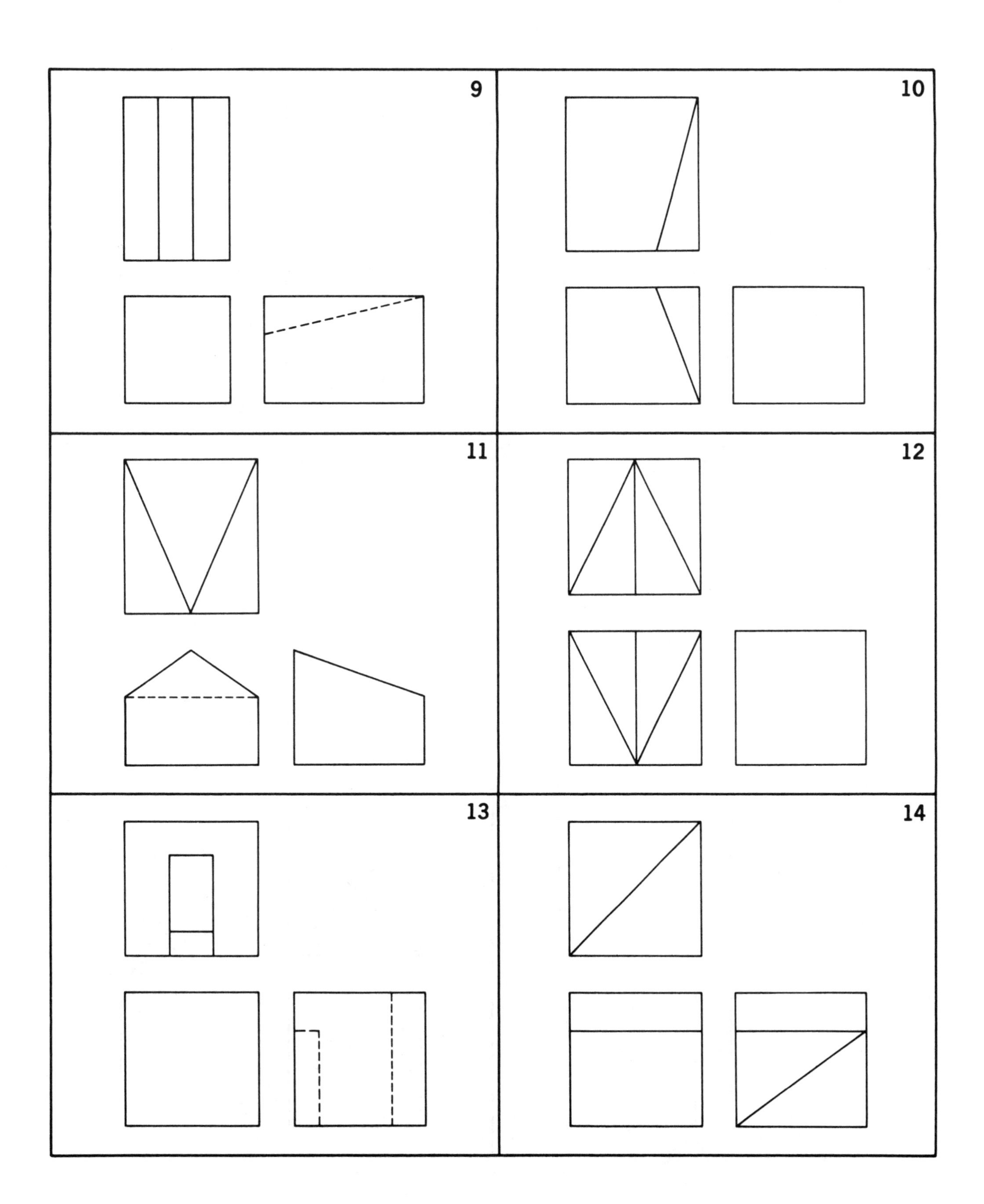
9
10
11
12
13
14

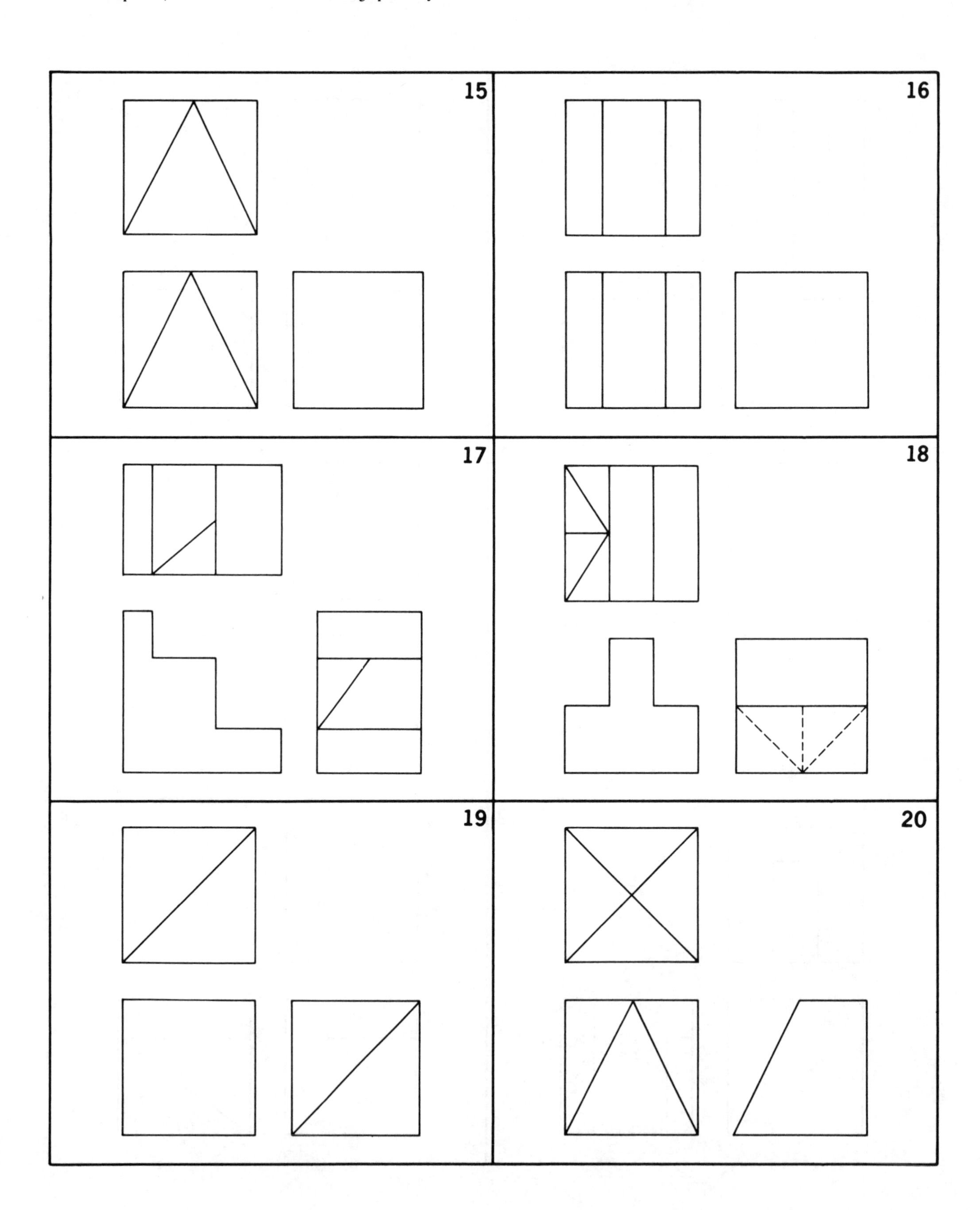
15
16
17
18
19
20

Part B (Missing Views)

Complete each set of drawings by sketching in the missing third view. The first problem is solved for you in gray. Answers are found in Appendix A.

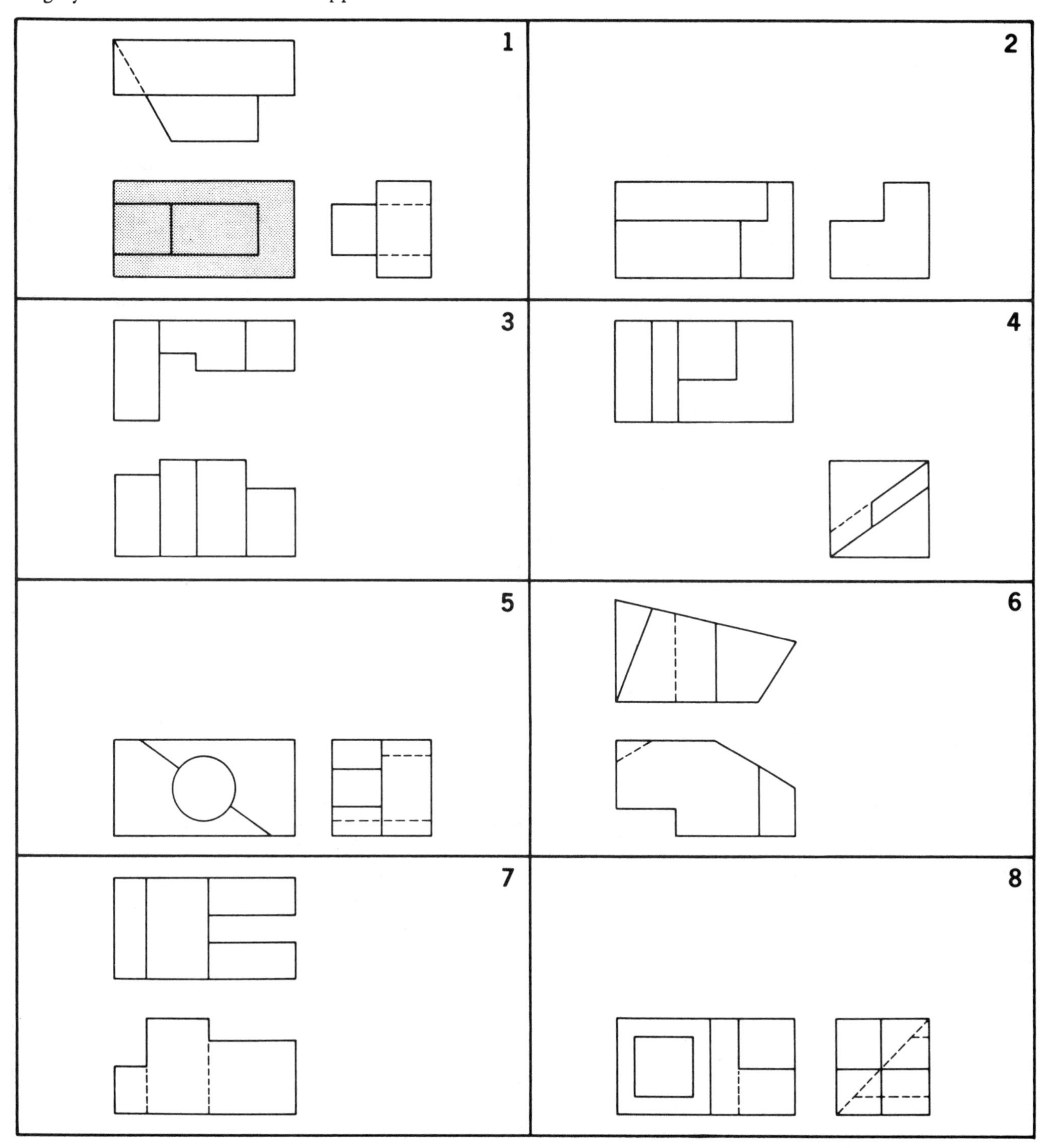

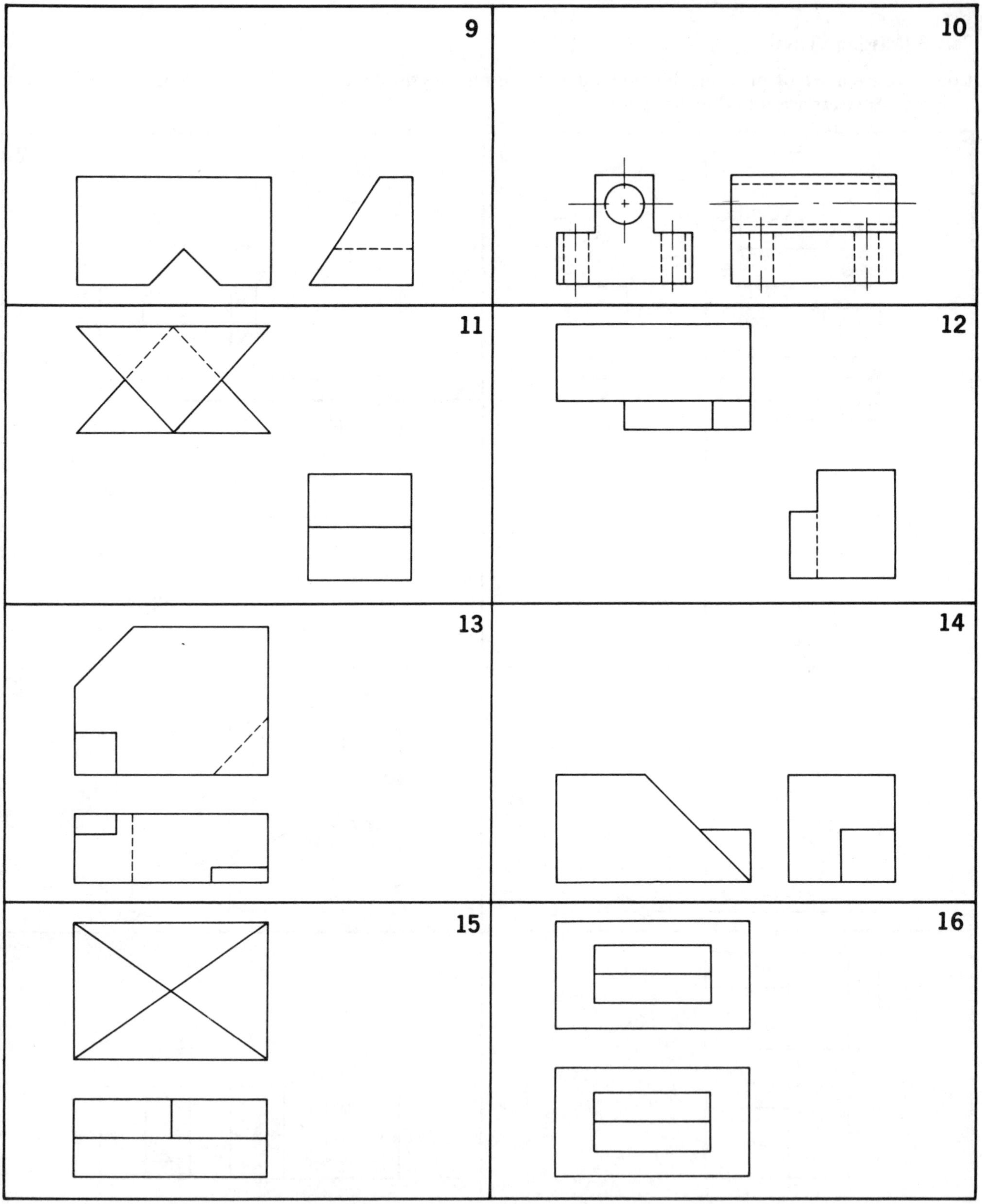

9

10

11

12

13

14

15

16

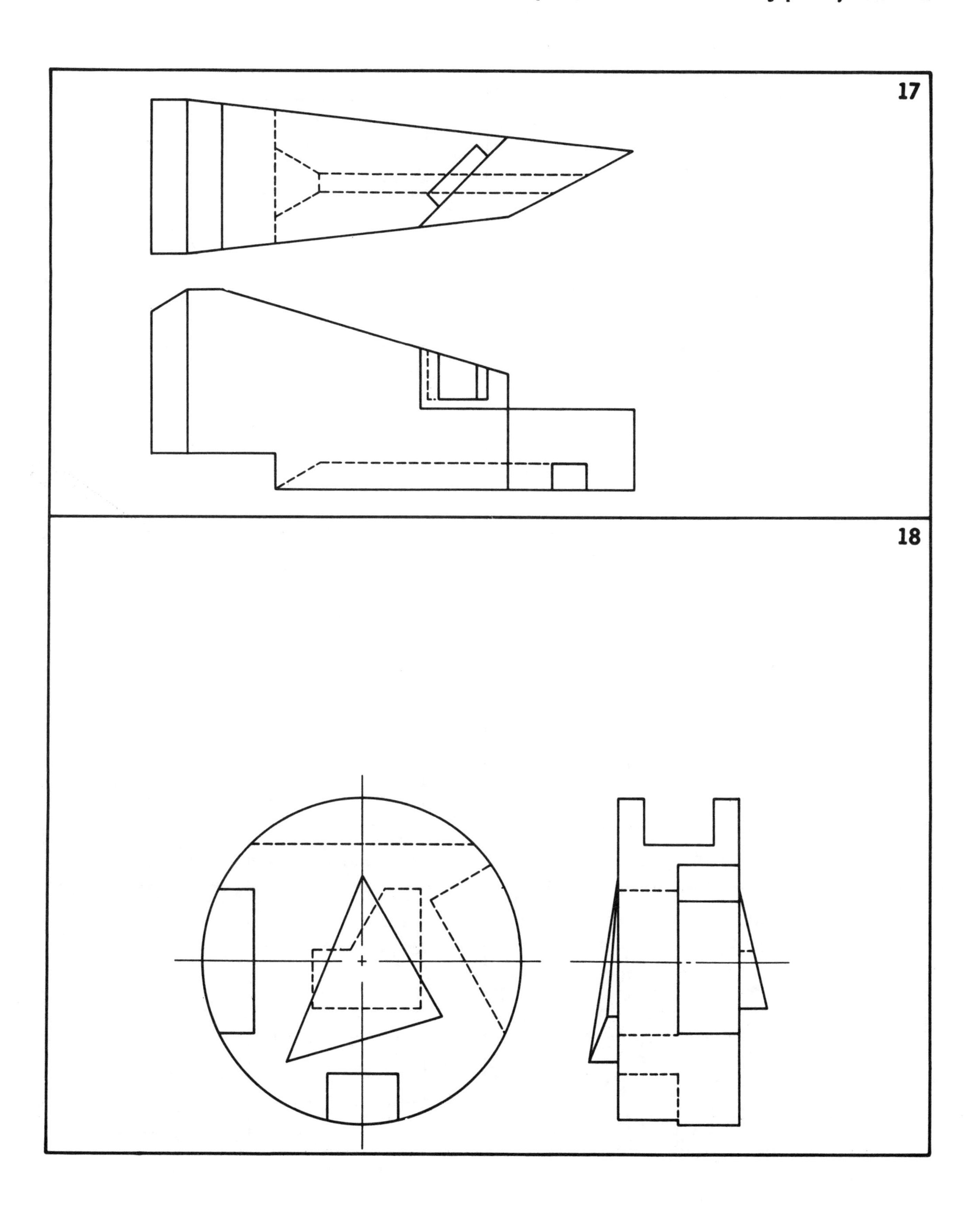
17
18

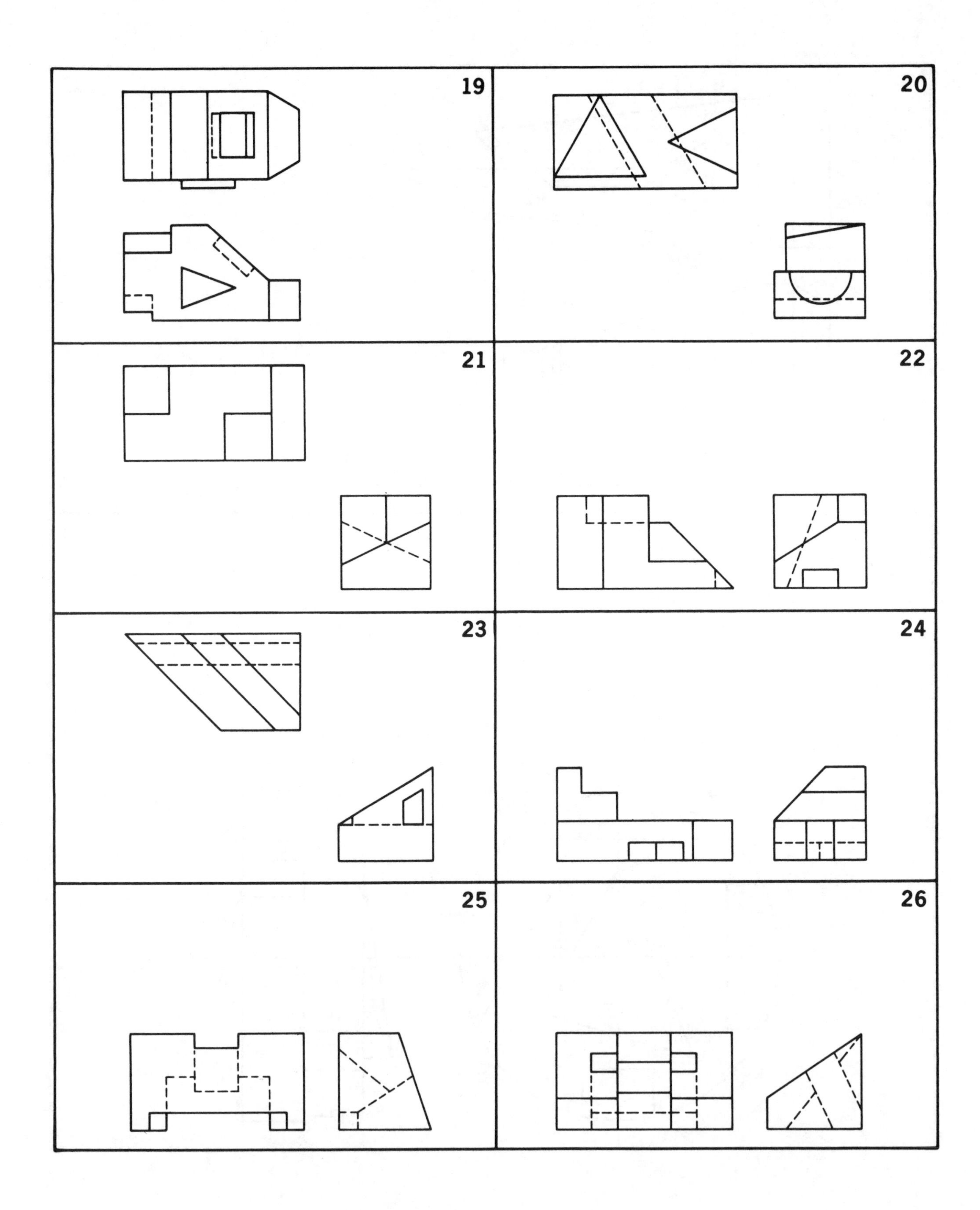
19
20
21
22
23
24
25
26

Part C (Line Identification)

Name the ten lines found on the sample drawing by writing the correct name of each in the space provided at the right.

1 ____________________ 6 ____________________
2 ____________________ 7 ____________________
3 ____________________ 8 ____________________
4 ____________________ 9 ____________________
5 ____________________ 10 ____________________

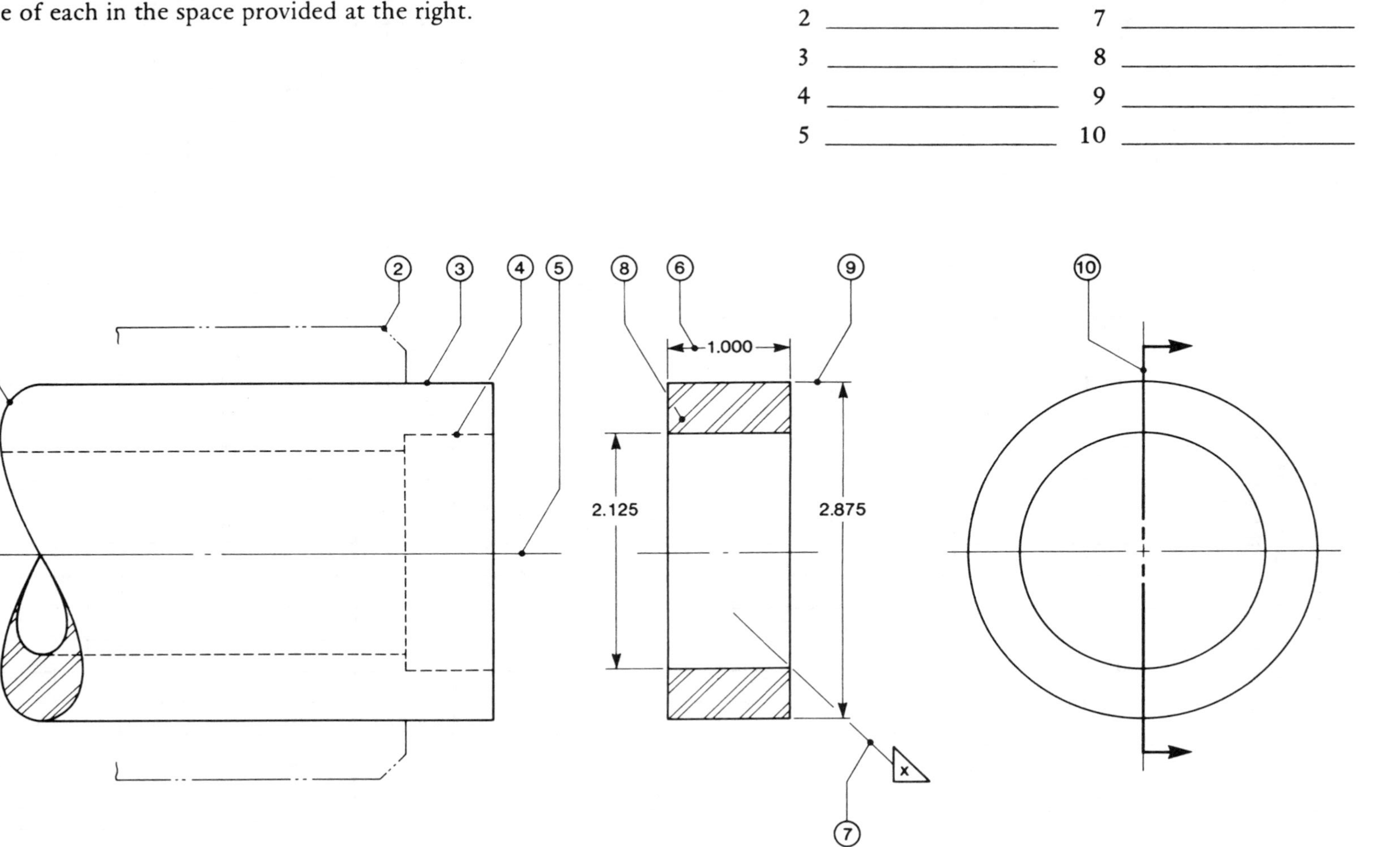

Part D (Developing an Orthographic Drawing)

Develop a three-view drawing for each of the following 18 exercises. Use the graph paper to draw your views. The front view is designated by the arrow. Answers are found in Appendix A.

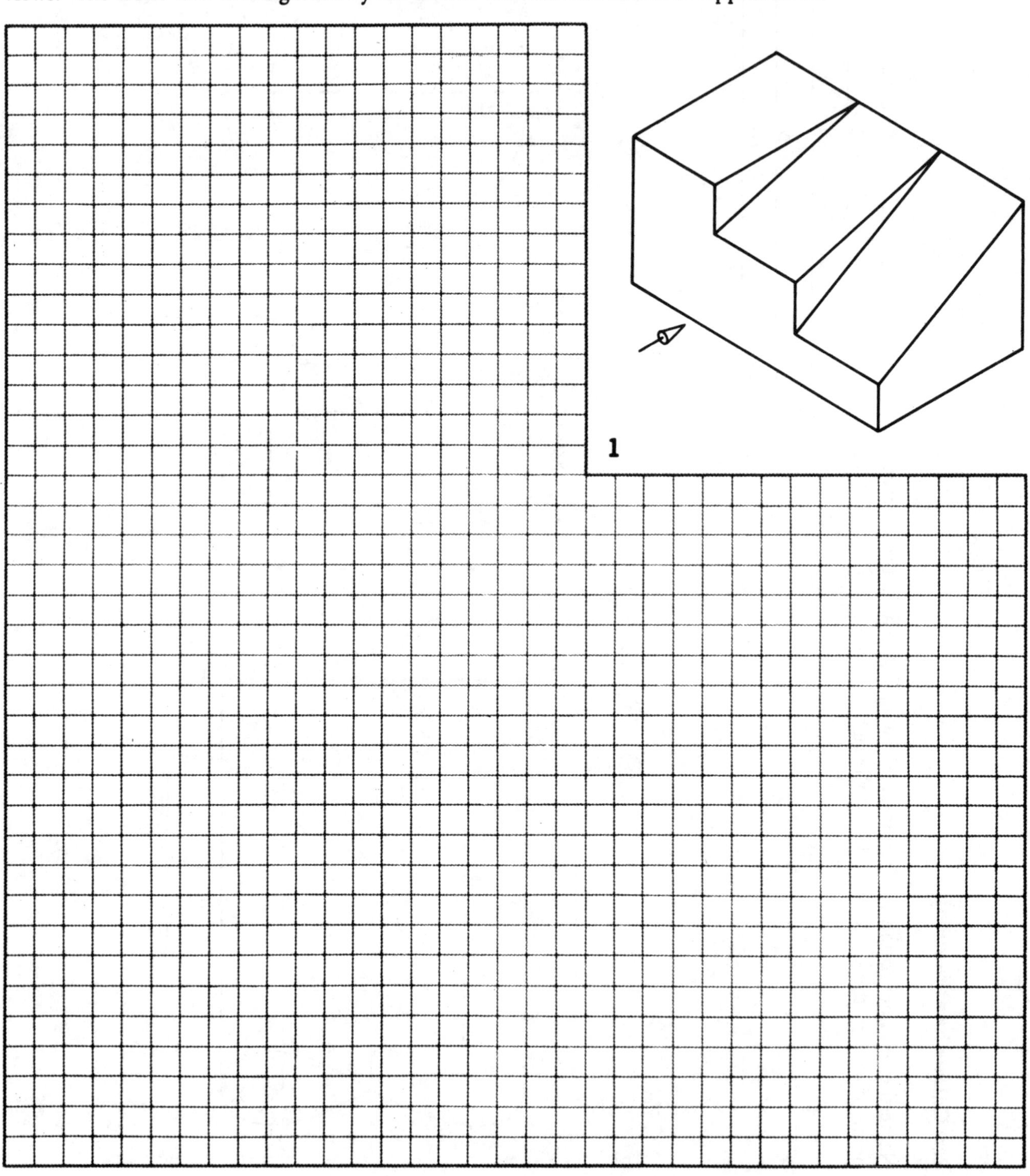

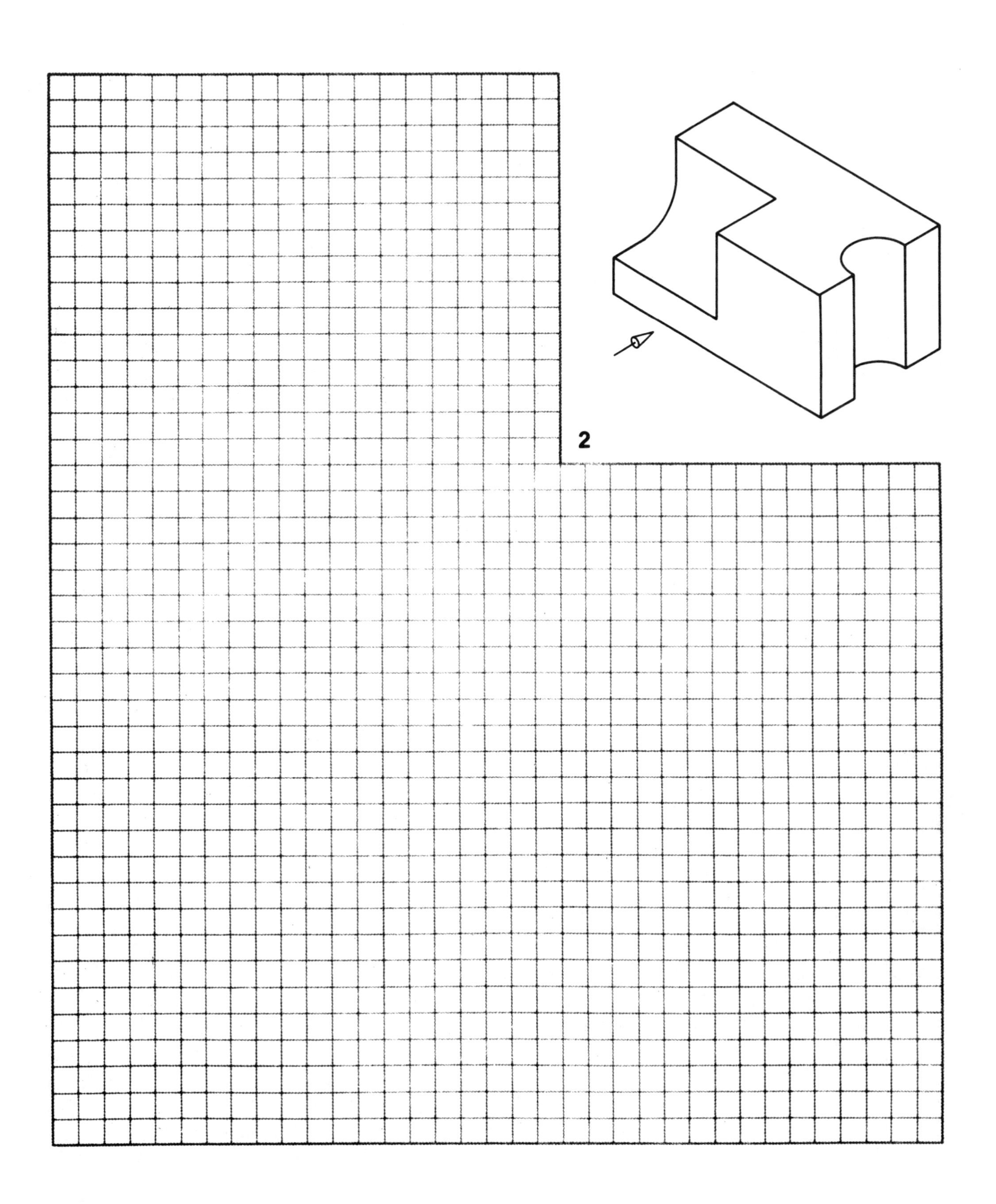
2

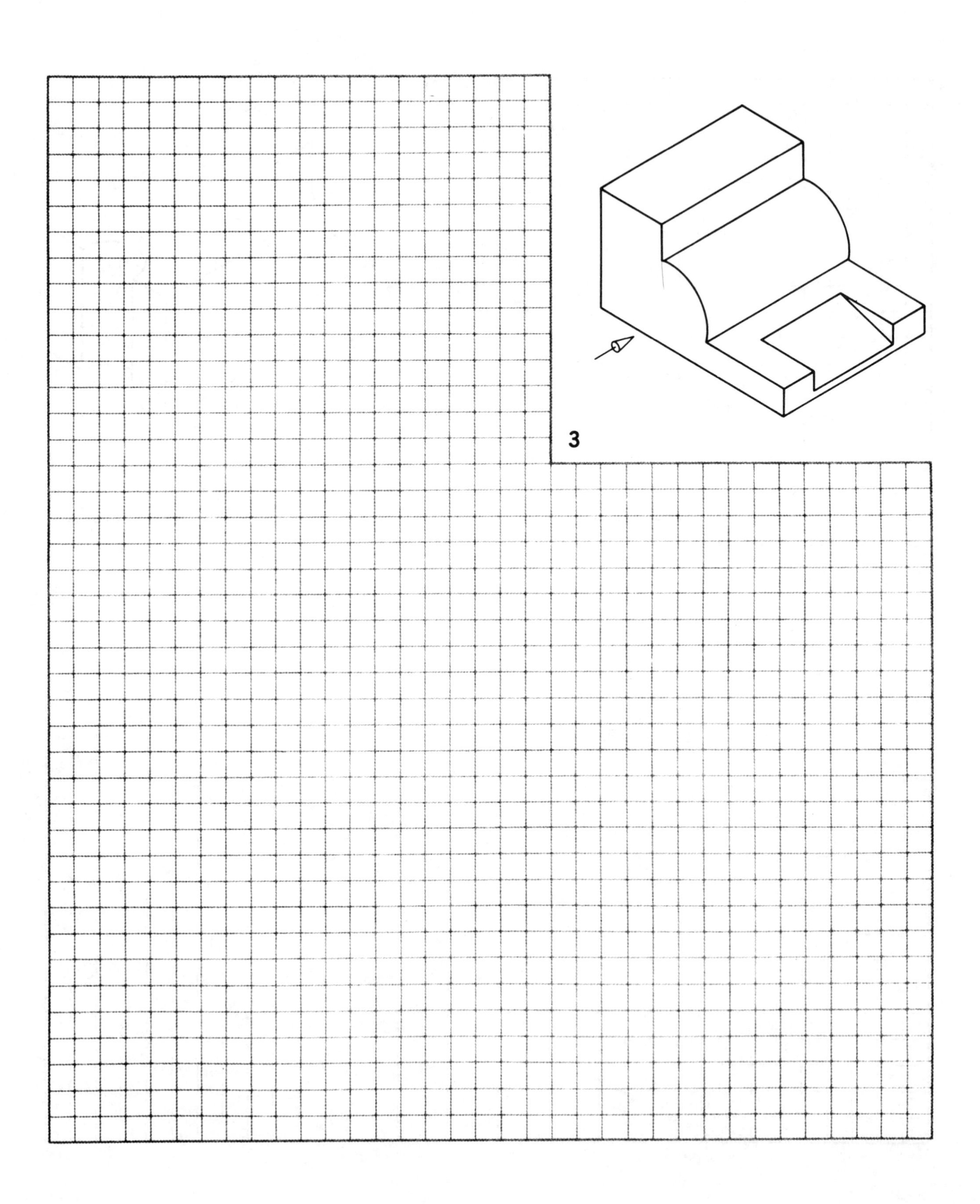
3

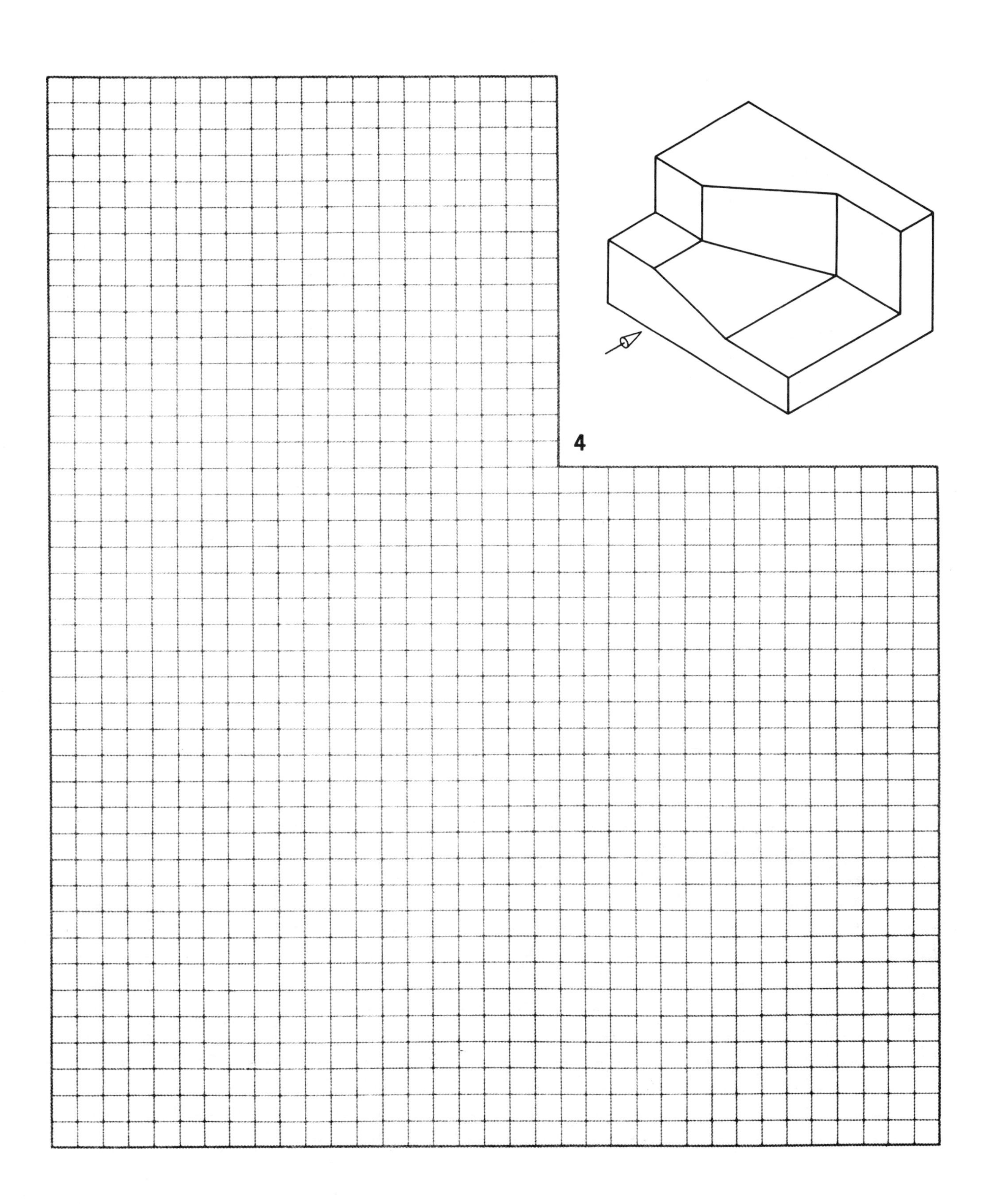
4

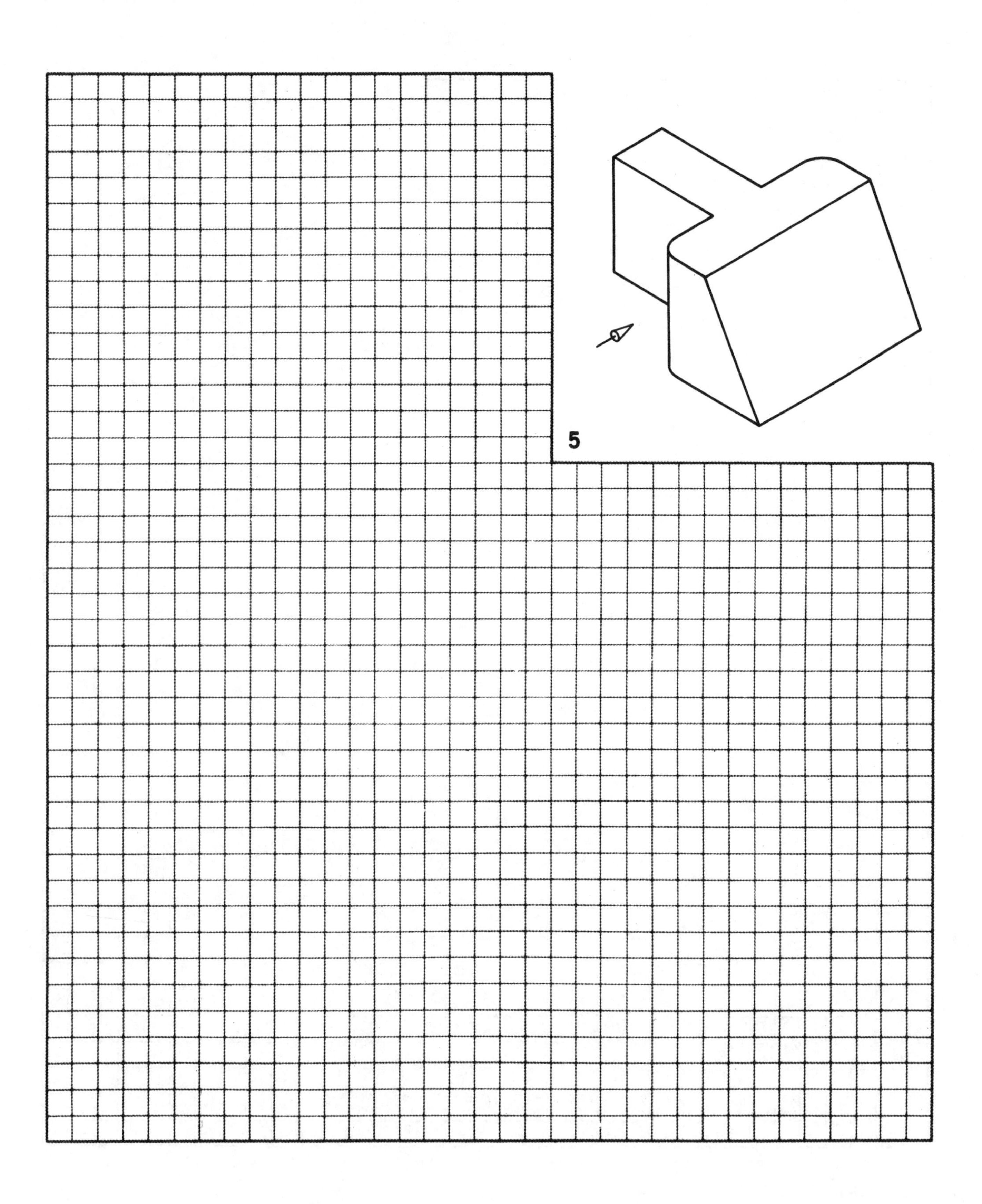
5

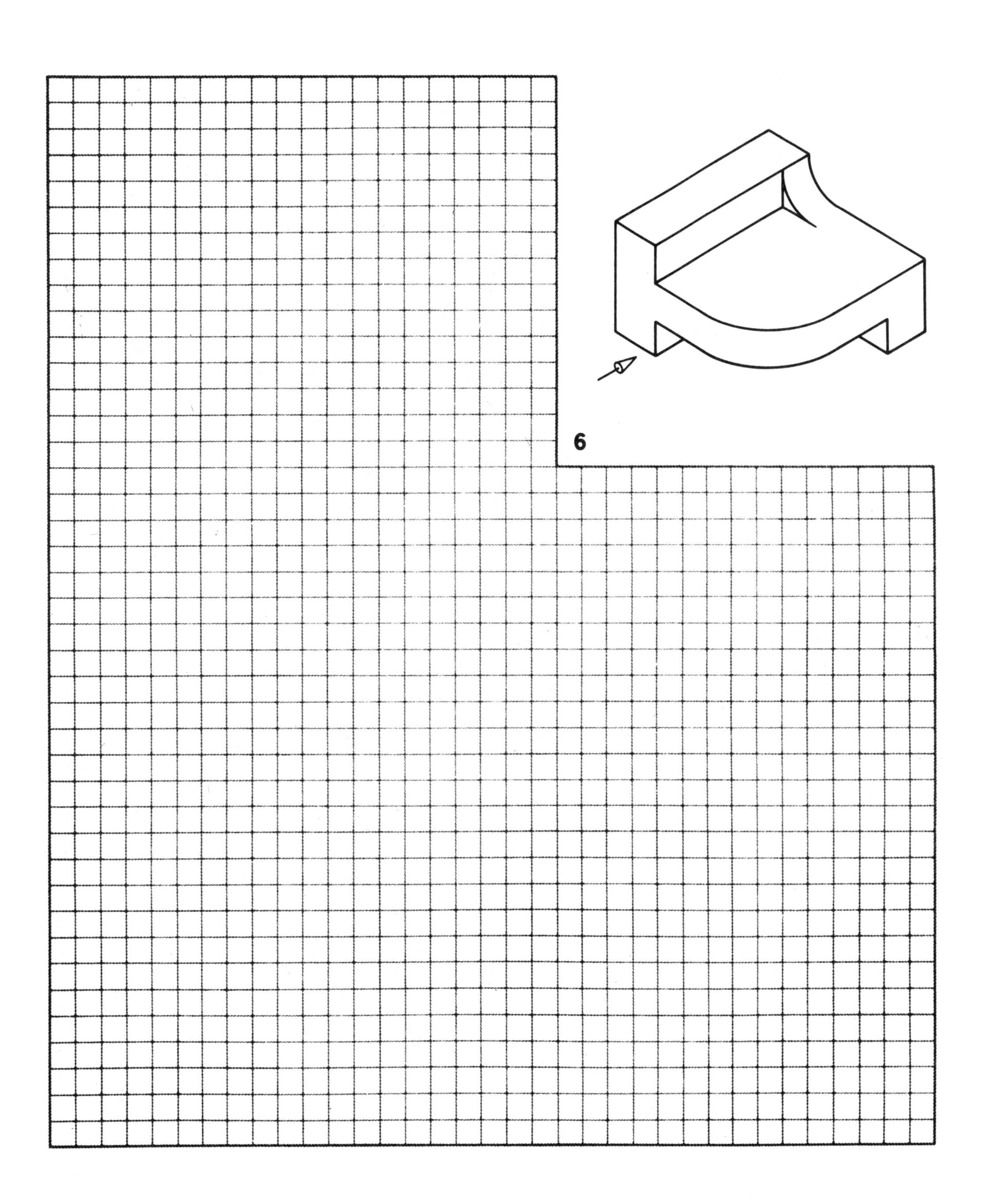
6

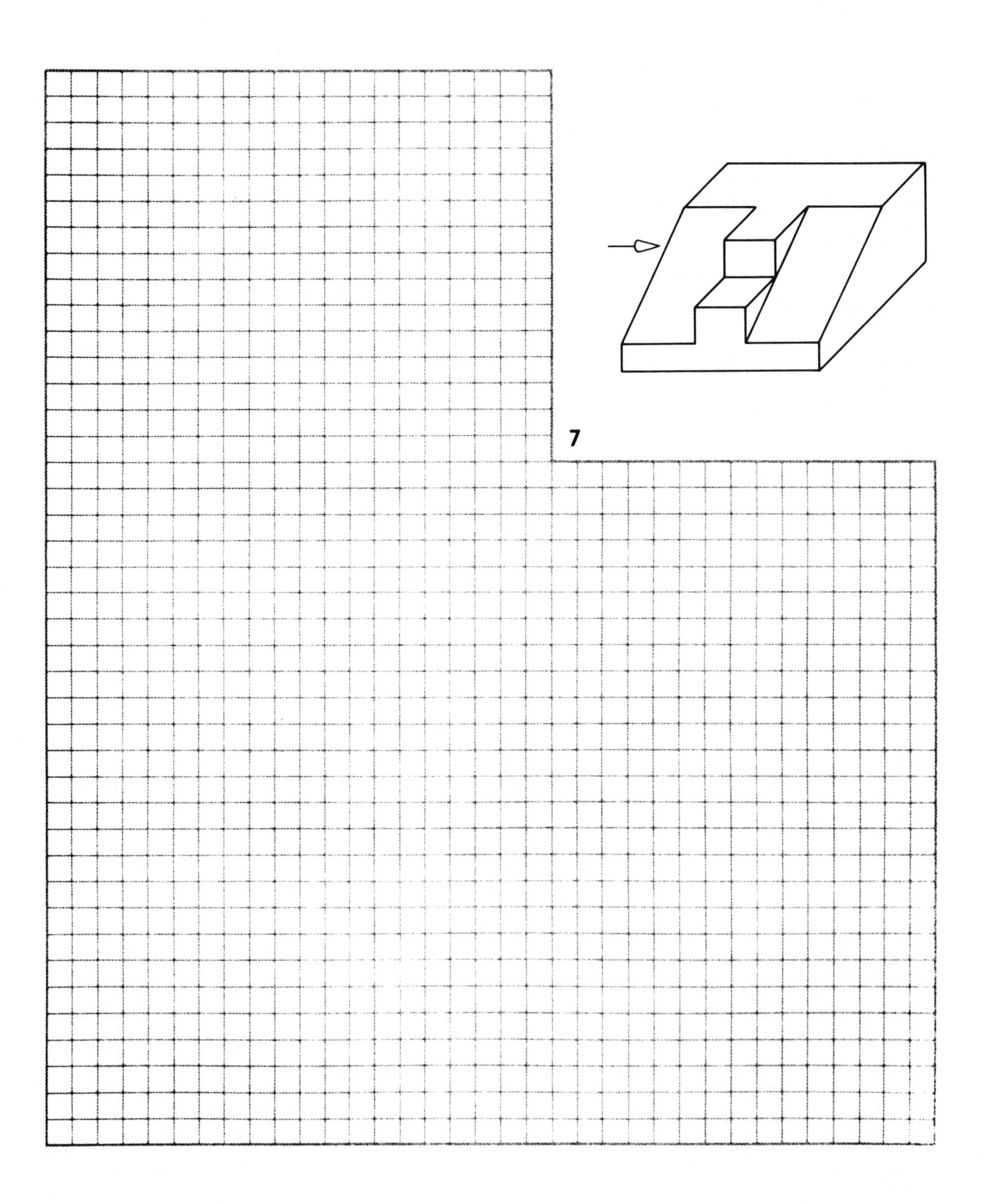
7

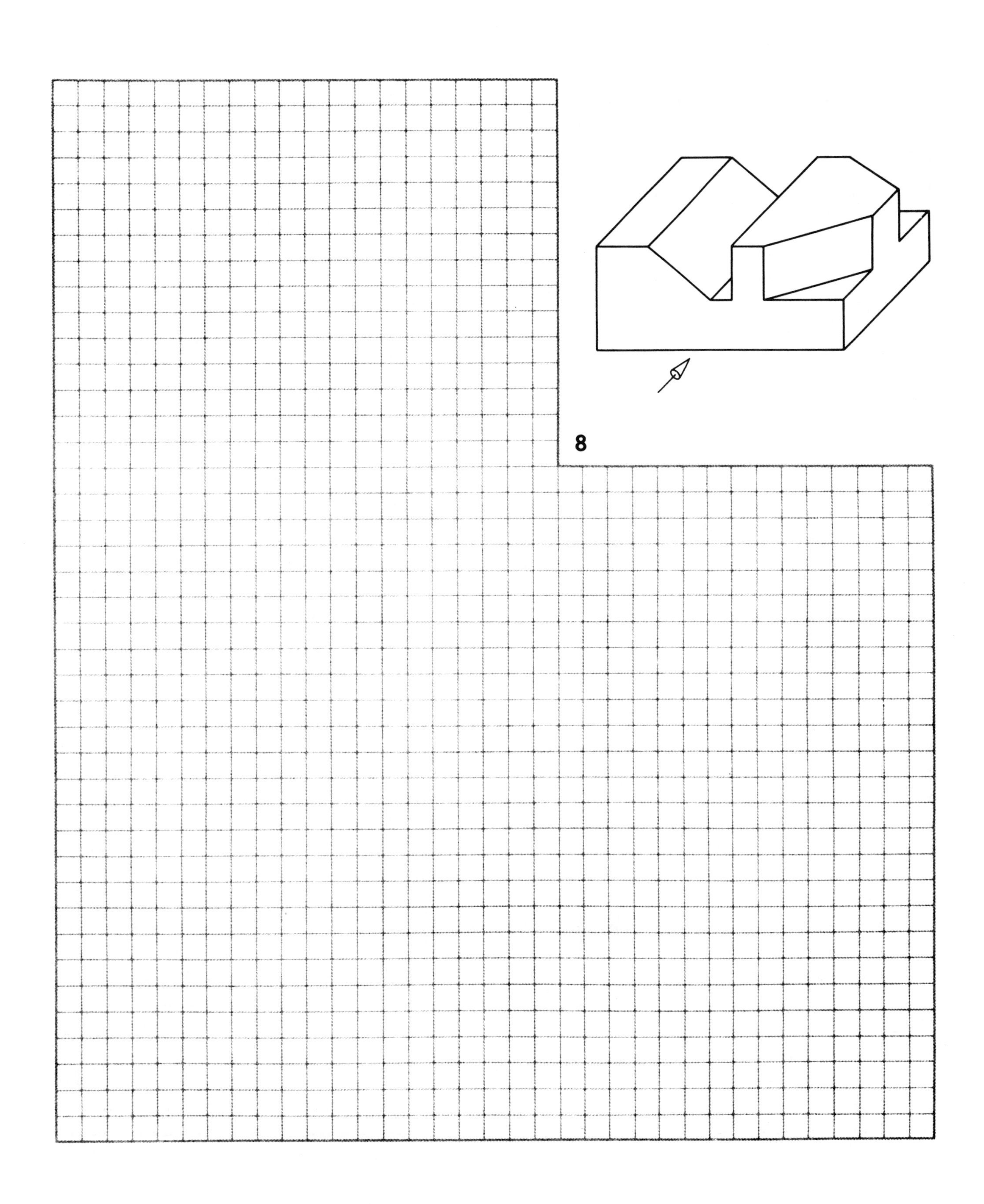
8

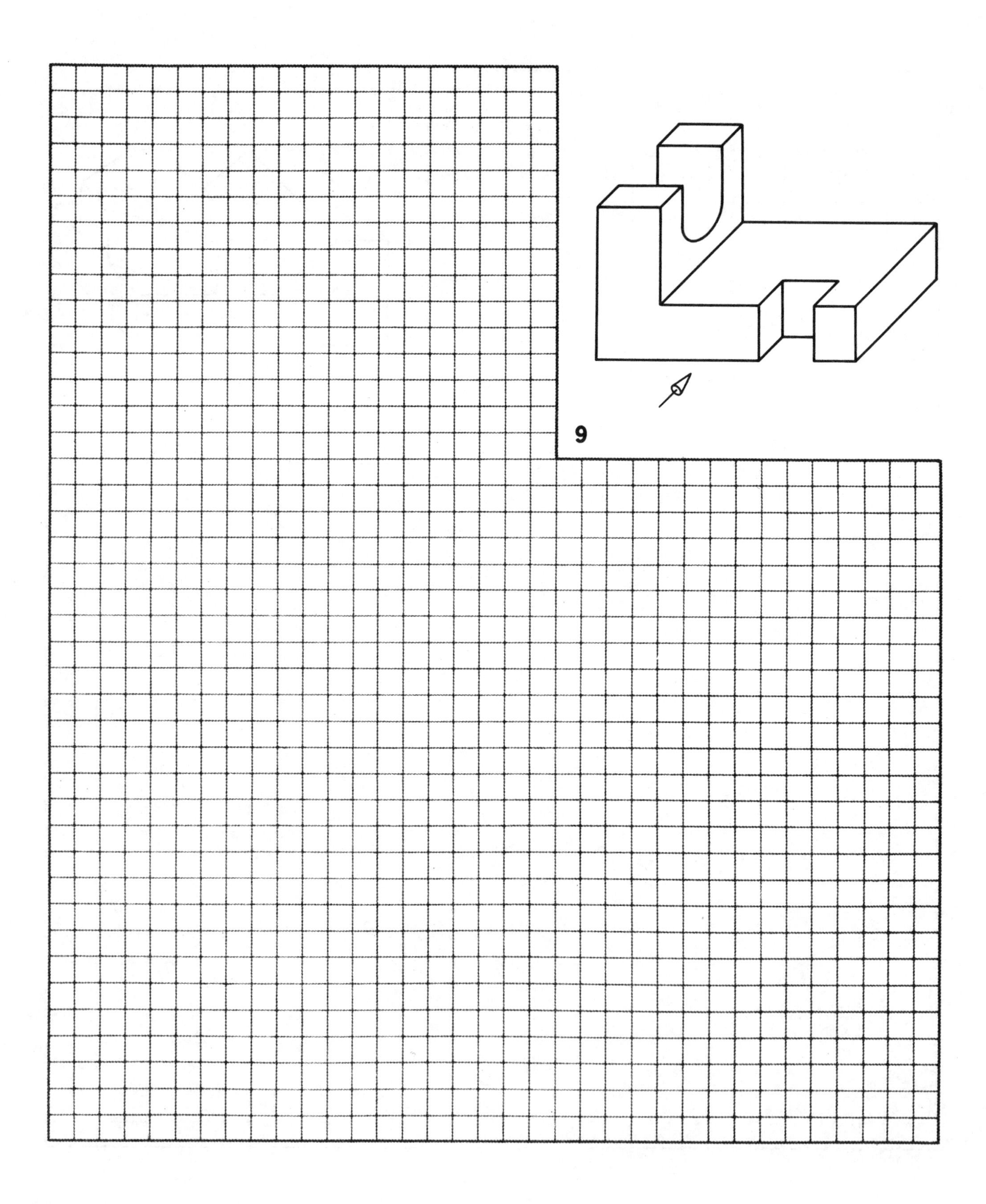
9

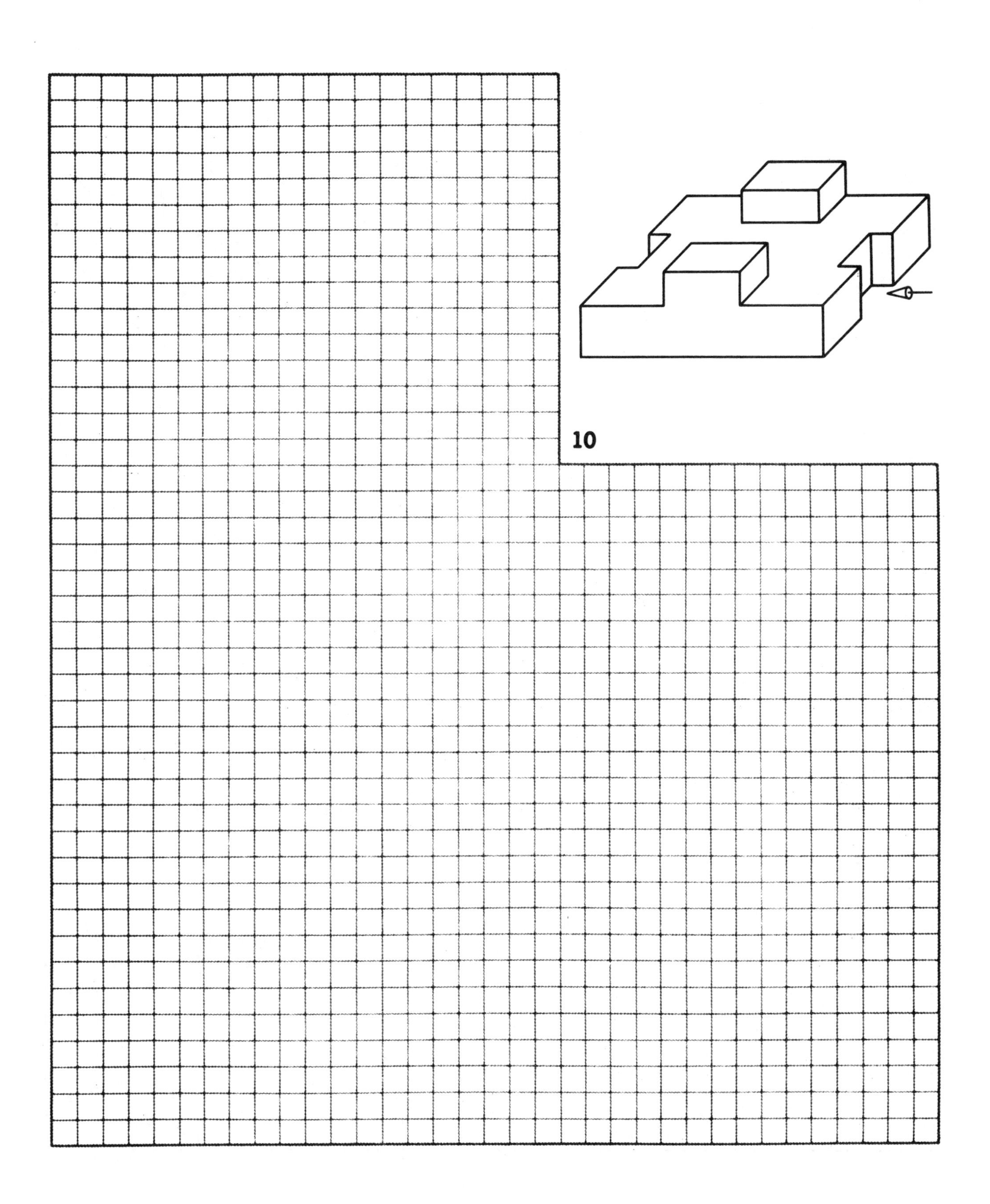
10

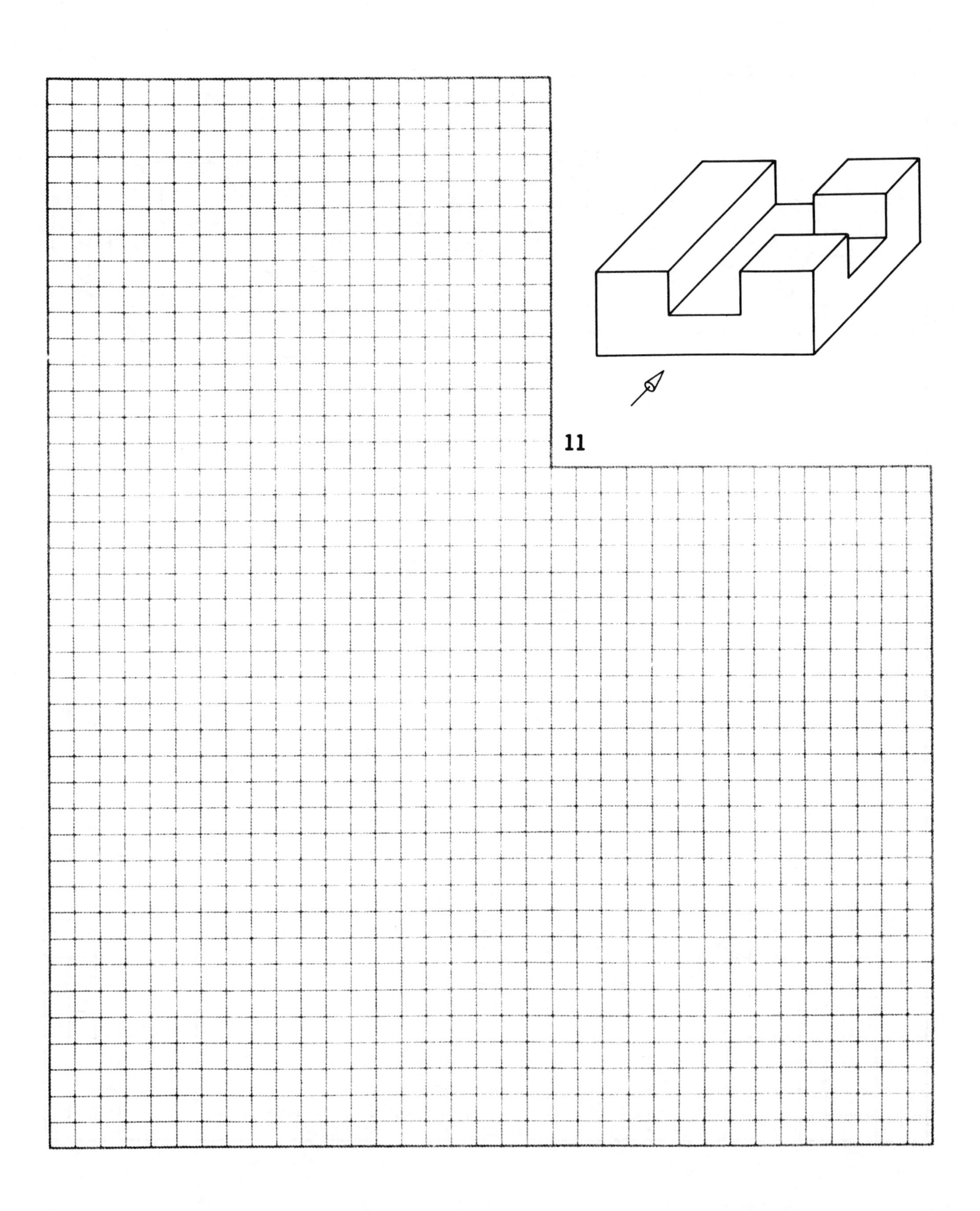
11

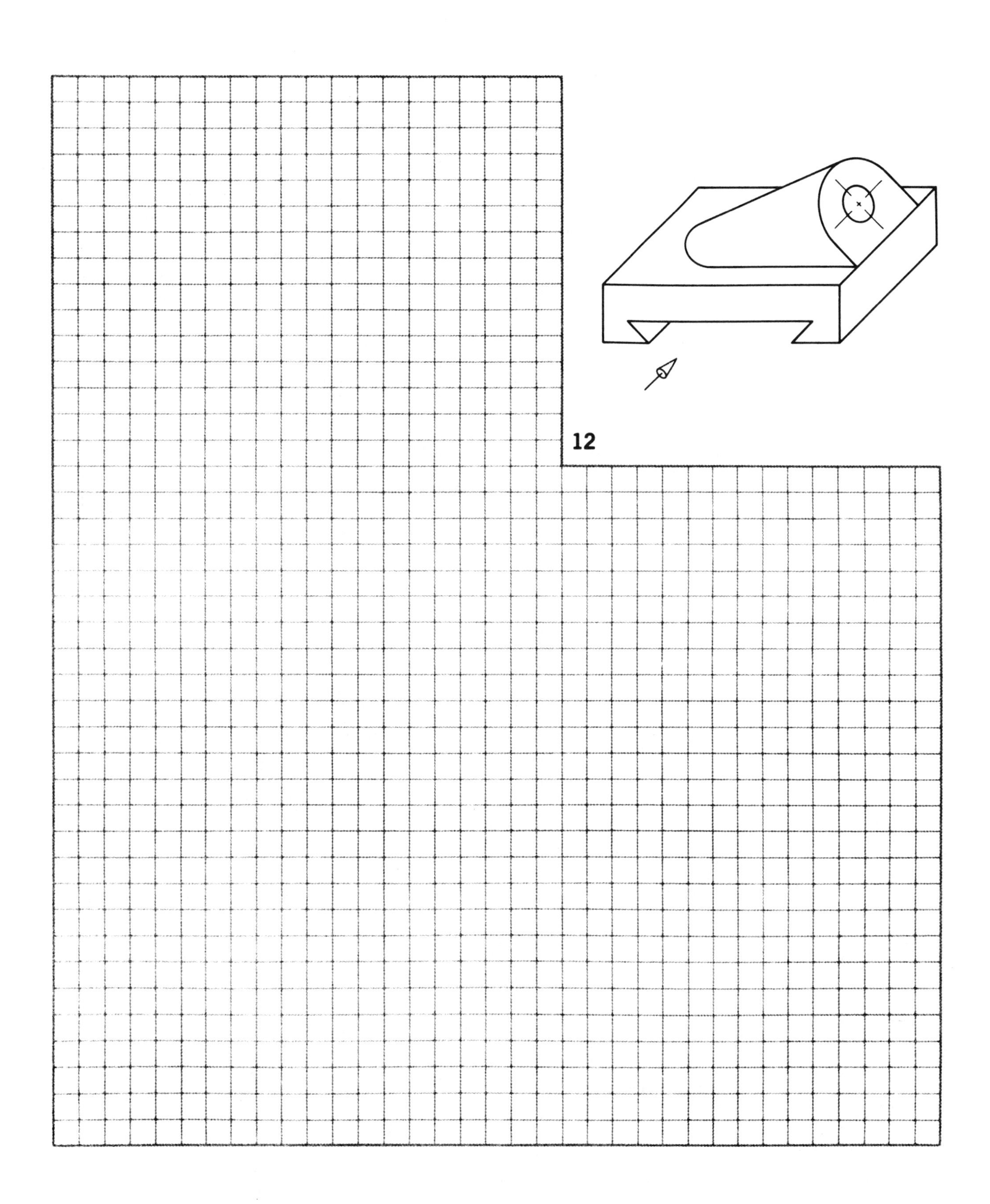
12

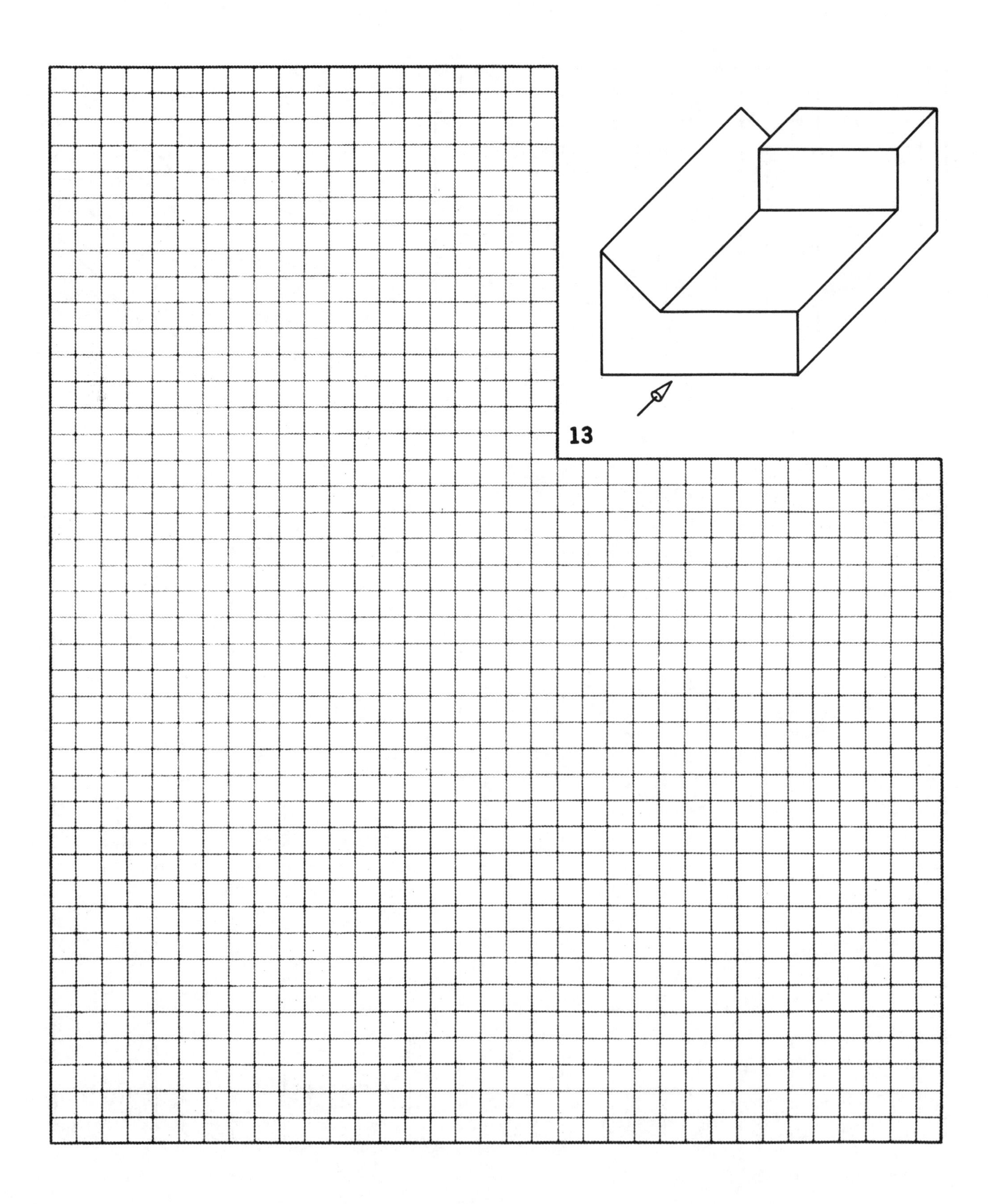
13

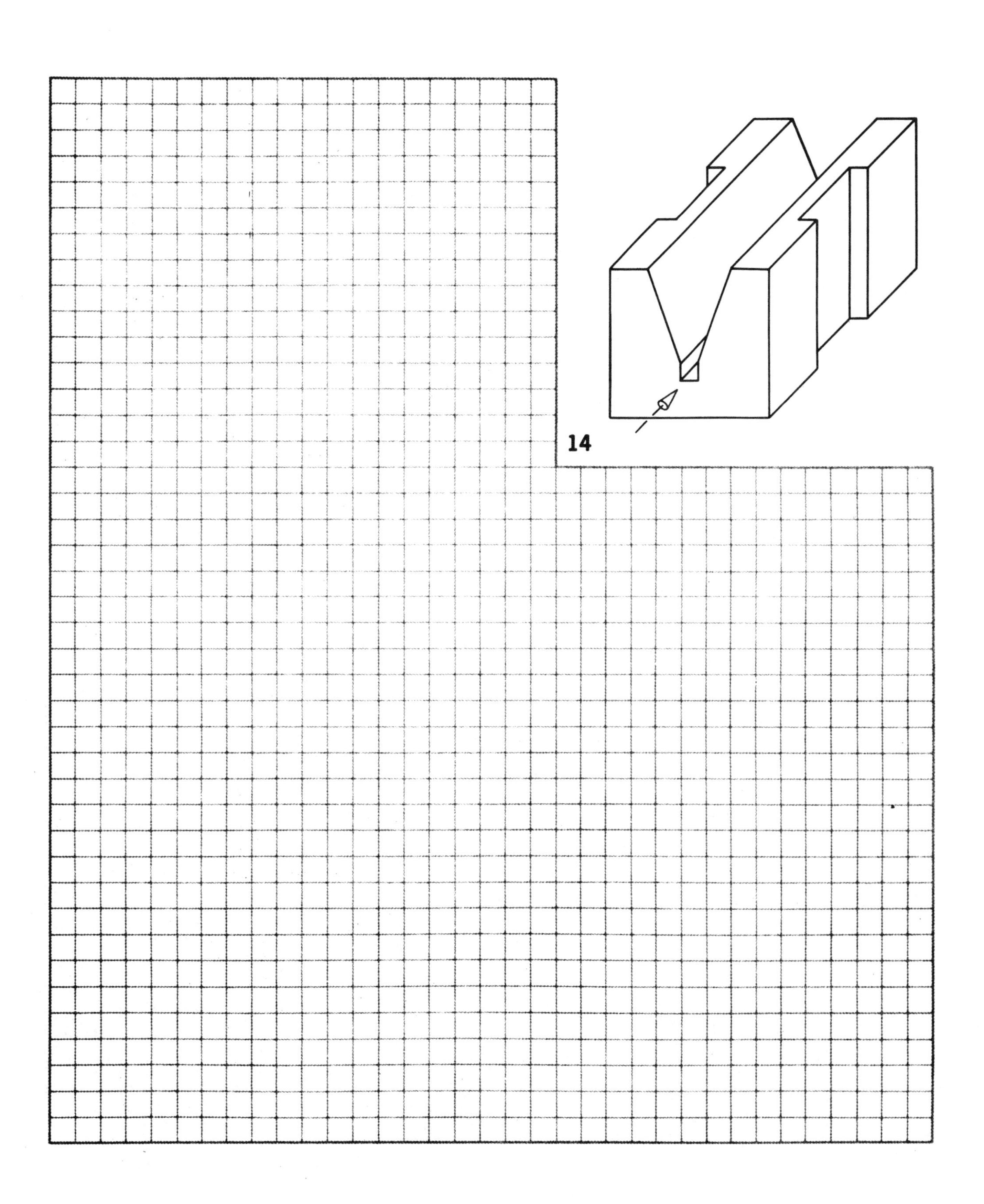
14

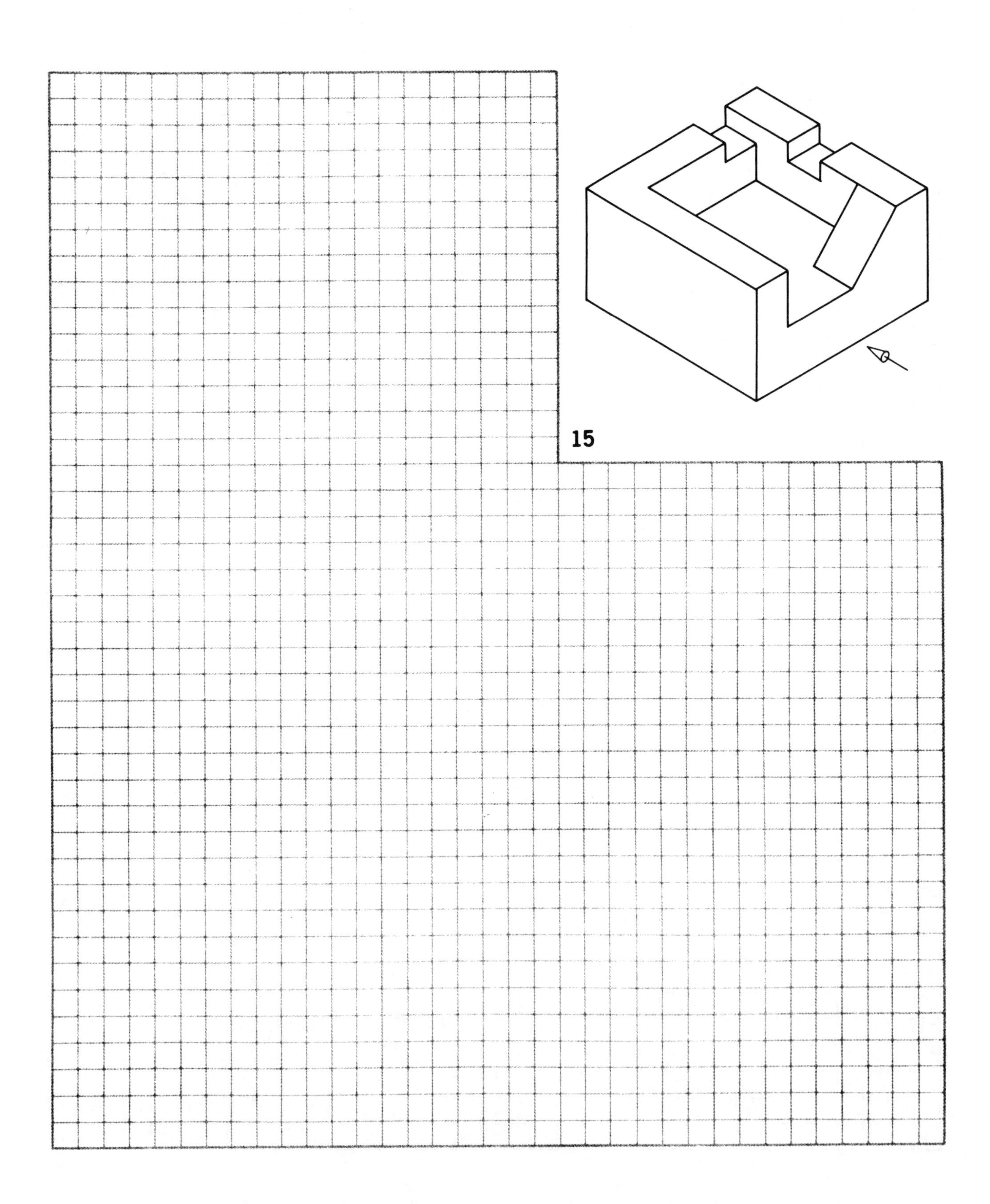
15

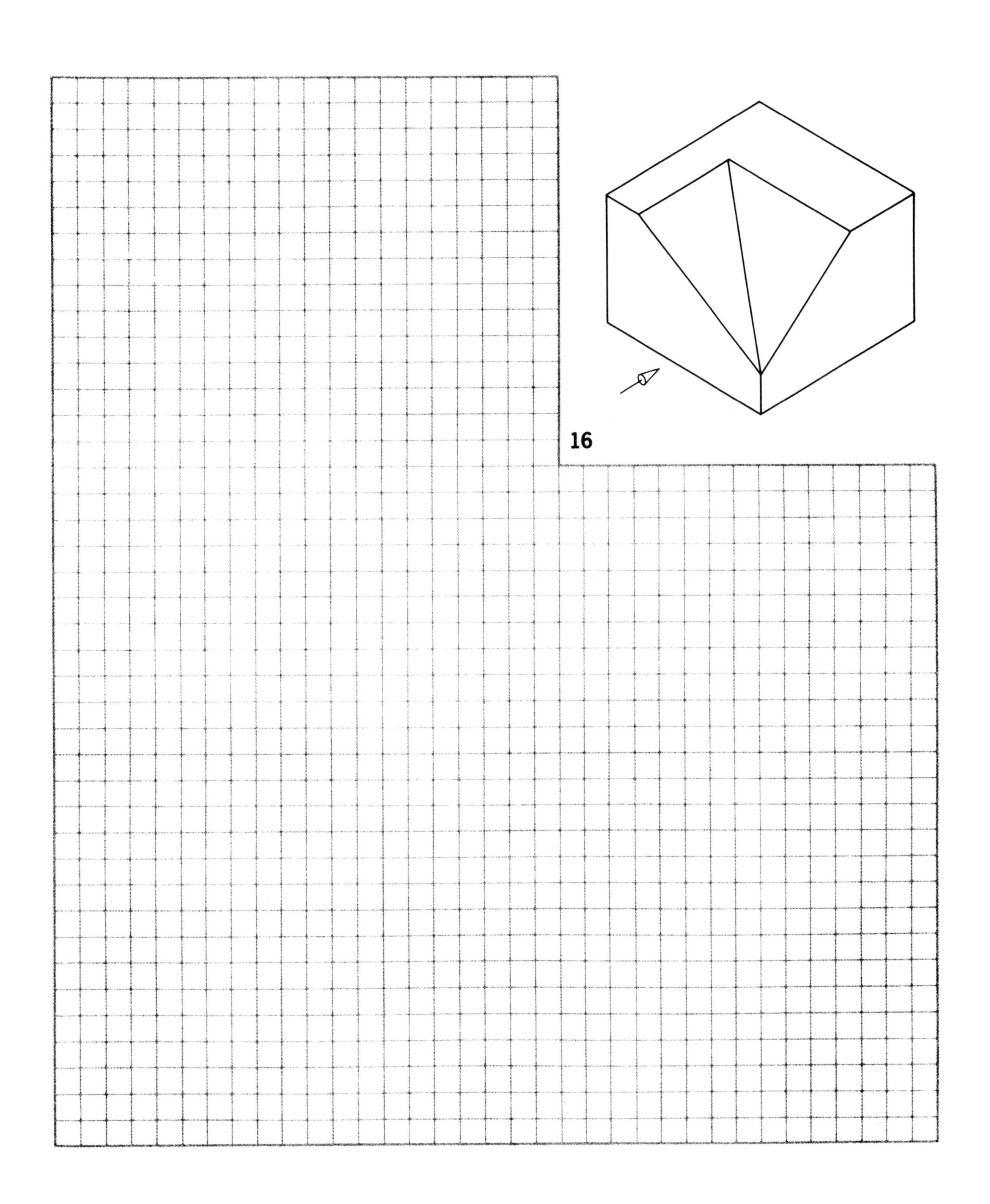
16

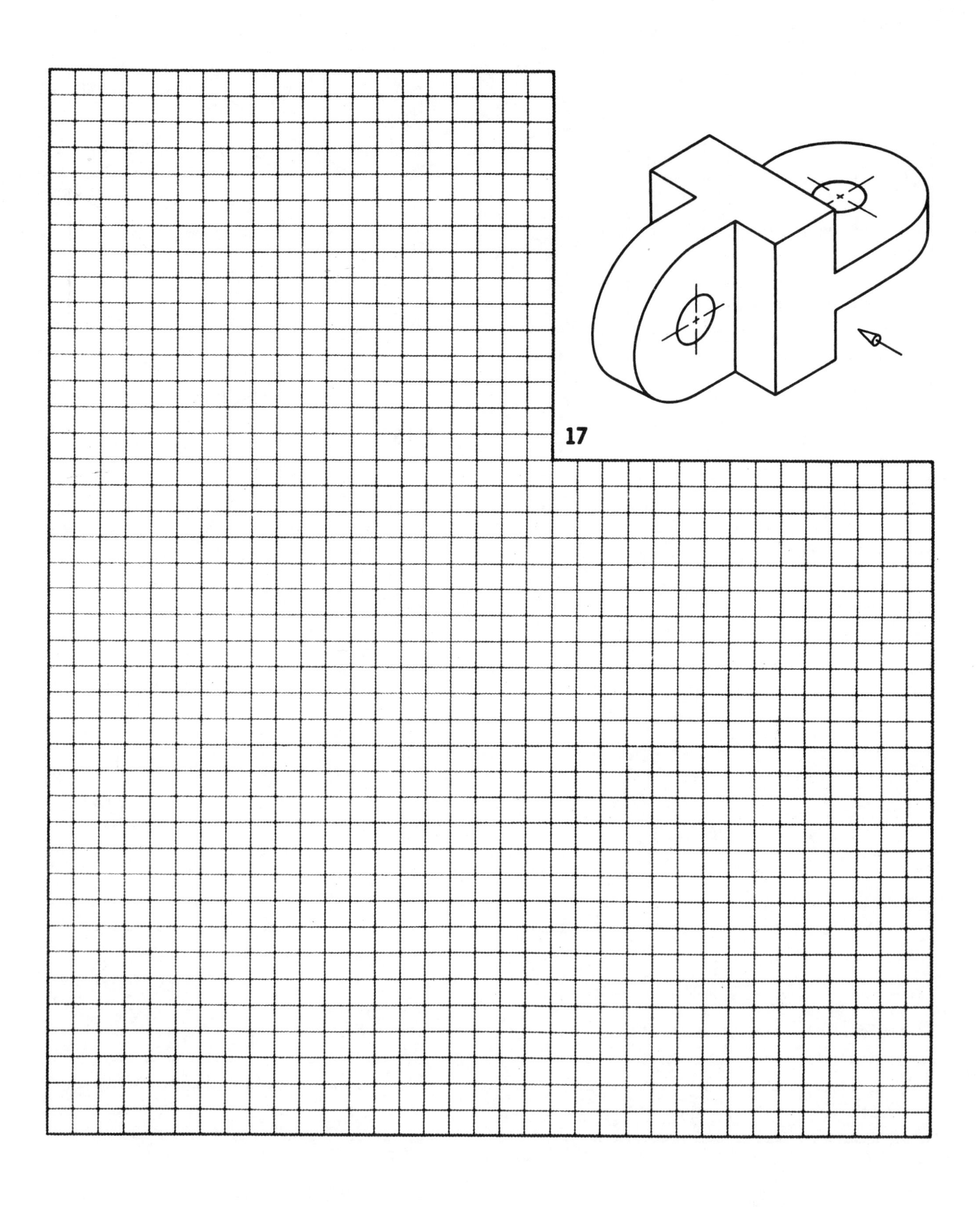
17

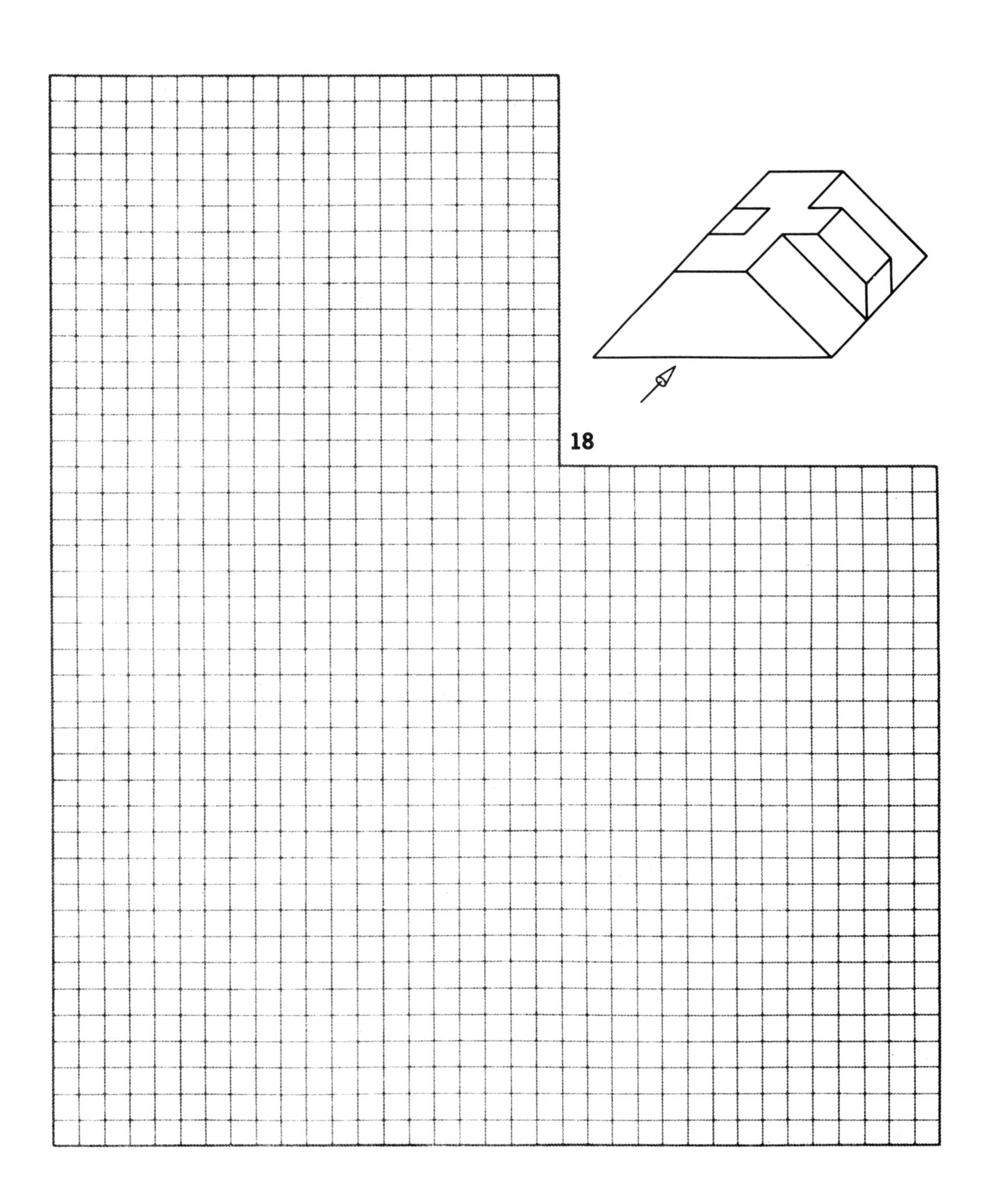
18

Chapter III

AUXILIARY VIEWS AND SECTIONAL VIEWS

Objects that are complex in shape, or have many interior features, can be difficult and confusing to depict in multi-view orthographic projection alone. Therefore, the engineer or draftsman uses two special techniques of engineering drawing to give you a clear picture of how such an object should be constructed. These techniques are called auxiliary views and sectional views.

AUXILIARY VIEWS

As you learned in Chapter II, the three regular multi-view planes of projection are the vertical, horizontal, and profile. However, some objects have one or more inclined (slanted) surfaces or other features that are not parallel to any of the three regular planes of projection. These surfaces and features, which are said to be *oblique* to one or more of the regular planes of projection, will appear foreshortened in one or more of the orthographic views of the object.

The basic rule of dimensioning requires that a line be dimensioned only in the view where its true length is projected, and that a surface with its details be dimensioned only in the view representing its true shape. Therefore, to satisfy this rule, we have to create an imaginary plane that is parallel with the inclined line or surface that we want to project in its true shape. This plane, which is not one of the regular planes, is called an *auxiliary plane* (see Fig. 3-1).

For example, look at the front view of Fig. 3-1 and note that it shows an inclined surface. However, in the right side view and the top view, the inclined surface appears foreshortened—not in its true shape or size. In a case like this, a projection called an *auxiliary view* is used to show the true shape and size of the inclined face of the object.

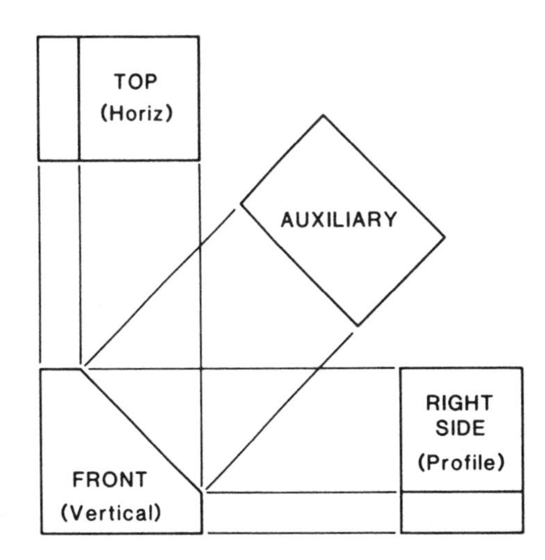

Figure 3-1 Three-view and auxiliary view arrangement.

Single Auxiliary View

A single auxiliary view is made by projecting the inclined surface in a plane parallel to it, and perpendicular to one of the regular planes. This enables the viewer to see the surface as it actually appears when viewed perpendicularly, or from a 90 deg angle. A single auxiliary view is called a *primary auxiliary view*.

Only a primary auxiliary view is required to

show the object in Fig. 3-1 in its true shape and dimensions, because the inclined surface is perpendicular to a regular plane of projection, and the front view in Fig. 3-1 shows the surface as a line, which can be used as the plane from which to project the auxiliary view. A primary auxiliary plane is always hinged to the regular plane to which it is perpendicular.

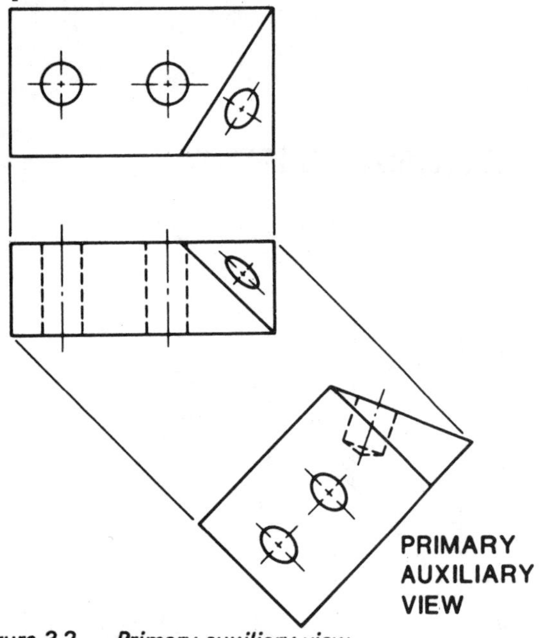

Figure 3-2 Primary auxiliary view.

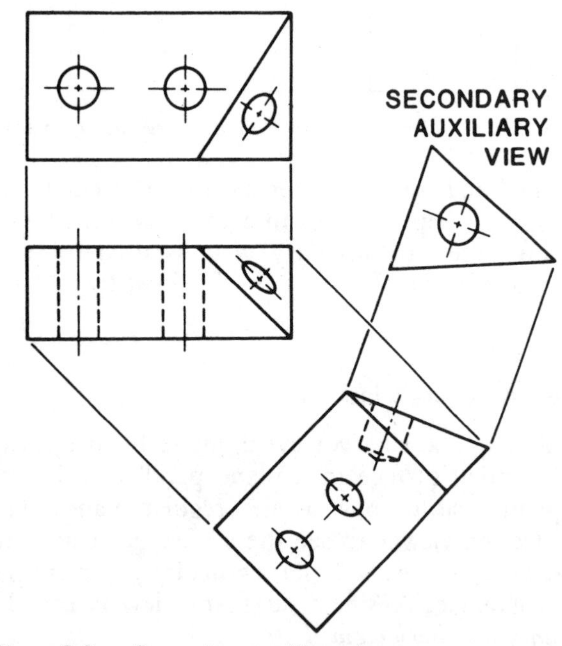

Figure 3-3 Secondary auxiliary view.

Double Auxiliary Views

Frequently, however, an object will have an inclined surface that is not perpendicular to any one of the regular planes of projection, and is not shown as a straight line on any of the regular views. When this is the case, a *primary auxiliary view,* in which the inclined surface appears as a straight line, must be projected first (see Fig. 3-2). Then, a *secondary auxiliary view,* in which the surface appears in its true dimensions, is projected from the primary, as shown in Fig. 3-3. The plane of the secondary auxiliary is oblique to all of the principal views, but is perpendicular to the primary auxiliary plane from which it is taken.

SECTIONAL VIEWS

Sectional views are used to illustrate the internal features of an object. In Chapter II, you learned that a thin dashed line—called a *hidden line*—on an engineering drawing or blueprint represents interior detail that cannot be seen. On complex parts, however, these hidden lines become so numerous that they create a confusing tangle. To avoid this tangle, draftsmen add sectional views to

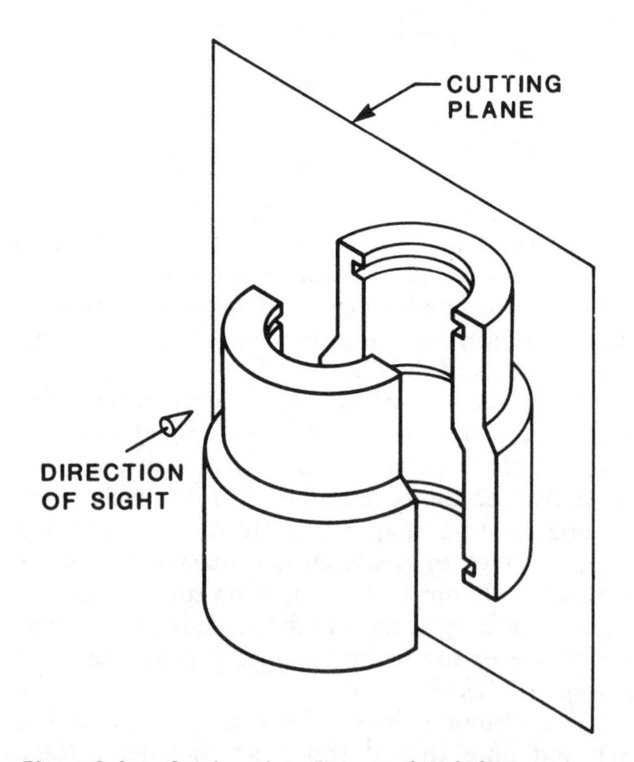

Figure 3-4 Cutting plane for a sectional view.

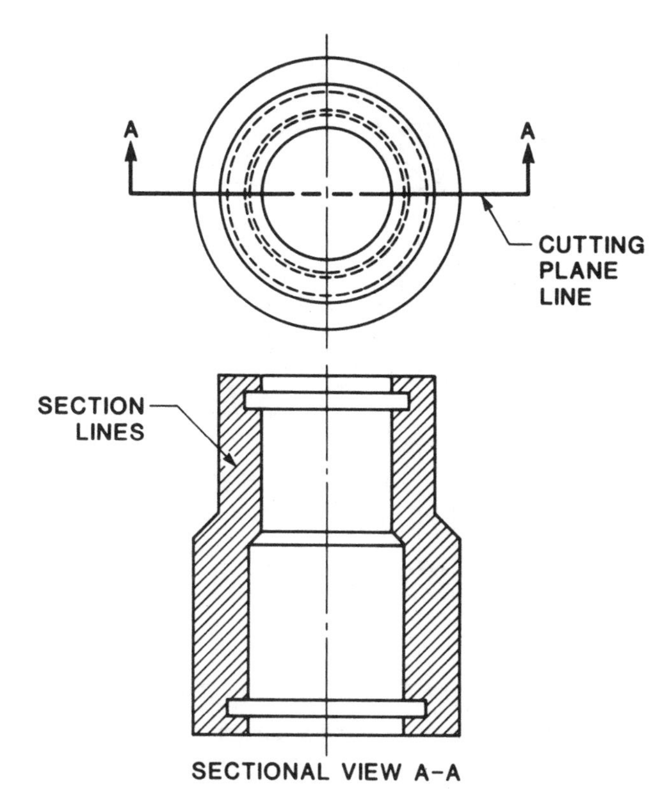

Figure 3-5 Exposed lines and surfaces of a sectional view.

the drawing which depict interior detail easily and clearly.

A sectional view is obtained by imagining that a portion of the object has been cut away—such as by sawing—to expose internal lines and surfaces. The path of the saw is considered to be the *cutting plane*, or the plane on which the cut was made, as shown in Fig. 3-4. When the sawed portion is removed, the formerly invisible lines of the object are exposed to view. These internal lines are then drawn as visible object lines, and the surfaces through which the cutting plane passes are cross-hatched to make the newly-visible portions stand out as shown in Fig. 3-5.

Sometimes a cutting plane line is added to the drawing to show the path of the cutting plane, and the direction of view is shown by arrows (see Fig. 5). The portion of the object toward which the arrows point is retained, and rotated orthographically to become another (sectional) view.

The three basic types of sectional views are: full, half, and broken or partial.

Full Sections

Full sectional views are obtained by passing the cutting plane across the entire object, exposing the whole inner surface, as shown in Fig. 3-6.

Half Sections

Half sectional views are obtained by passing two cutting planes at right angles to each other along the centerlines or symmetrical axes of the object, exposing one-half of the inner surface, as shown in Fig. 3-7. The object is shown as if it had been cut along the horizontal and vertical center-

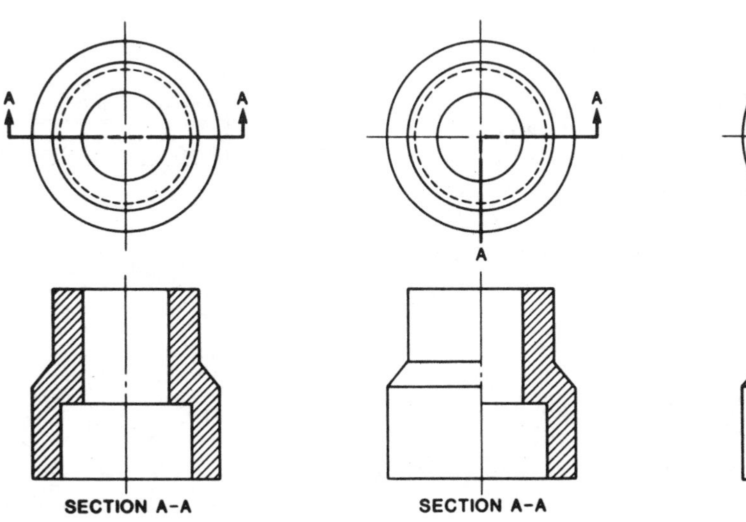

Figure 3-6 Full sectional view.

Figure 3-7 Half sectional view.

Figure 3-8 Broken-out or partial sectional view.

lines with either one-fourth or three-fourths removed. Usually, the unexposed portion of the sectional view does not show invisible object lines, but it may if the internal area is very complex.

Broken-Out or Partial Sections

Sometimes it is necessary to show a single detail, or a closely-related group of details, that exist within the interior of an object. If these details are all that are needed, then the broken-out or partial sectional view is employed. The usual cutting plane line is eliminated in the adjacent view, and the detail in the broken-out view is bounded by an irregular break line as shown in Fig. 3-8.

Rotated or Revolved Sections

In addition to the three basic types of sectional views, there are other types of sections that you should recognize to understand the full meaning of a drawing or blueprint. A rotated or revolved section is made directly on one of the principal exterior outline views of an object, such as bars, webs, ribs, etc. The cutting plane passes perpendicularly to the centerline or axis of the part to be sectioned, and the resulting section is revolved 90 deg into the plane of the paper. The sectional view is either superimposed on the object, or set between break lines as shown in Fig. 3-9.

Removed or Detail Sections

Removed or detail sections are similar to rotated sections and serve the same purpose, but are detached from the object and positionally located by a centerline. The centerline extends through the object and the section removed, as in Fig. 3-10A. By positioning the section view apart from the object, it can be enlarged to show greater detail (see Fig. 3-10B)

Auxiliary Sections

A sectional view that is related to an auxiliary projection is called an auxiliary section. Auxiliary sections conform to the same rules as auxiliary projections (see first part of this chapter). They are usually in the form of full sections, half sections, or broken-out (partial) sections.

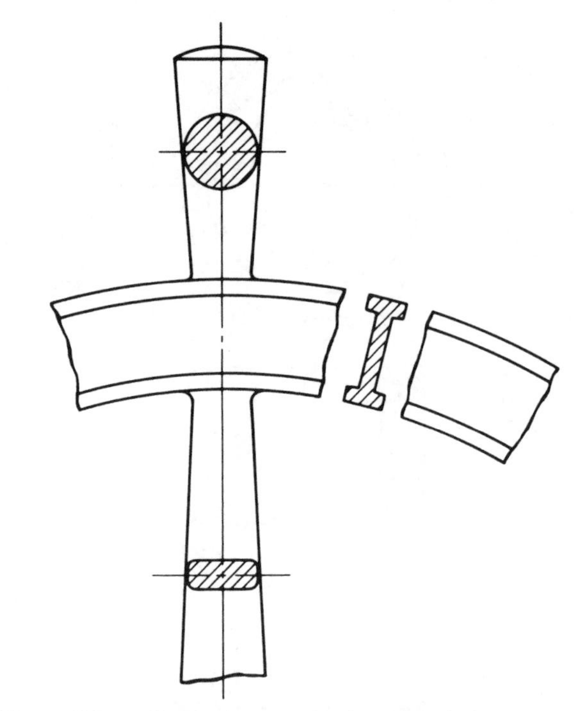

Figure 3-9 Rotated or revolved sectional view.

Assembly Sections

Assembly sections, as the name implies, are made up of a combination of parts. All of the types of sections described in the foregoing paragraphs may be used to make assembly drawings more clear and easier to read. The cutting plane for an assembly section is often offset to reveal the separate parts of a machine or object. Because the

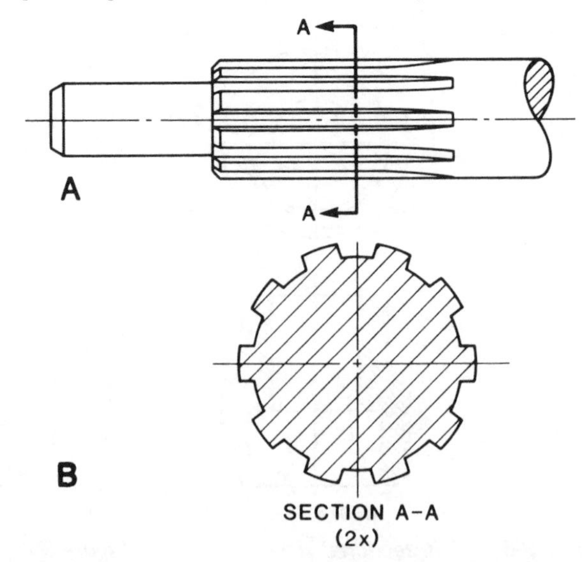

Figure 3-10 Removed or detail sectional view.

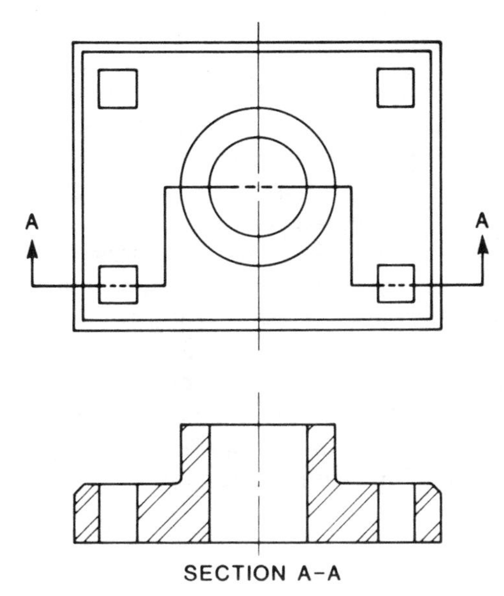

Figure 3-11 *Offset sectional view.*

separate parts do not need to be completely described, only those hidden details required for part identification or dimensioning are shown. Also, the small amount of clearance between mating or moving parts is not shown, because it would have to be greatly exaggerated and would confuse the drawing. Even the clearance between a bolt and its hole, which may be as much as 1/16 in., is rarely shown.

Offset Sections

In some cases, it is necessary to draw a series of two or more cutting planes in different directions passing through the object. When the cutting plane consists of two or more intersecting planes, the resulting view is called an offset section, as shown in Fig. 3-11. Reference letters may be used at the points where the cutting planes change direction, but are usually omitted if no confusion will result. The lines of intersection (where the cutting planes intersect) are not shown on the sectional view. The direction or position of each cutting plane is indicated by the cutting plane line.

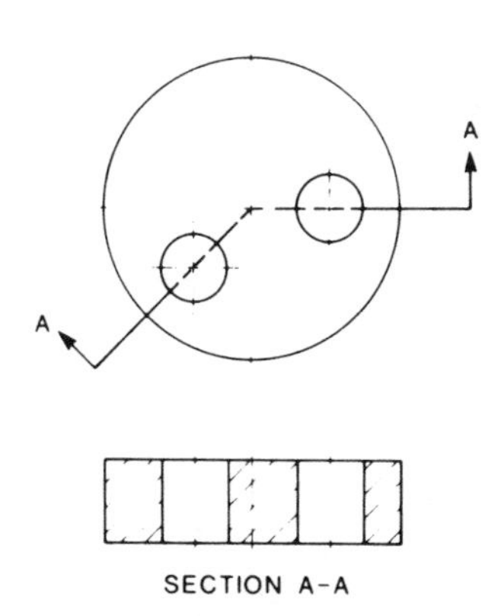

Figure 3-12 *Angular sectional view.*

Angular Sections

Angular sections show a part that is cut at an angle other than 90 or 180 deg, with the section revolved back to 180 deg so that details can be shown in their true dimensions. Figure 3-12 shows an example of an angular section.

Sectioning Rules

In preparing engineering drawings, the draftsman is guided by certain rules or conventions when he draws a sectional view. Your understanding of these rules will help you to read drawings quickly and accurately.

(1) Thin sections, such as sheet metal, structural shapes, packing gaskets, etc., are shown solid (not crosshatched) in sectional views, with a very narrow space left between the thicknesses of such parts.

(2) When a part is sectioned in more than one place, the spacing and direction of the crosshatching is the same in all sectioned areas.

(3) When adjacent parts are shown in sectional views, their crosshatchings are in opposite directions.

(4) When three adjacent parts are in section, two of them will have 45 deg crosshatching in opposite directions.

(5) If crosshatching using standard angles would appear nearly parallel to an object line of the part, different angles are used.

(6) Invisible object lines and details behind the cutting plane are not shown on sectional views unless they are needed to clarify the drawing.

(7) When a cutting plane passes through a rib, web, or similar parallel portion of the object, the crosshatching is omitted from those portions. In such cases, the cutting plane is considered to pass just in front of the rib or web.

Sectional Symbols

As you learned in Chapter II, the crosshatching most commonly used in sectional views is actually the standard graphic symbol for cast iron. This symbol is often used, regardless of the actual material of the object. However, various standard symbols are available to represent other materials (see Fig. 2-1J). These symbols are not meant to give exact descriptions of materials, but to serve as aids when you are reading a drawing.

SUMMARY

Auxiliary views show an inclined surface or line in its true size and shape. This makes it easier to visualize the object, and—at the same time—allows an otherwise oblique surface to be dimensioned according to standard drawing practices. Primary auxiliaries are projected from any of the six main orthographic views. Secondary auxiliaries are projected from primary auxiliary views.

Sectional views reveal an object's inner detail by graphically removing portions of the surface. This is done in a standardized way by using a cutting plane line or break lines. Either way, the purpose of sectional views is to simplify the drawing by eliminating hidden lines.

TRAINING PRACTICE

Blueprint No. 6

1. Draw Section A-A at a scale of 4 to 1.

2. Draw Section B-B at a scale of 2 to 1.

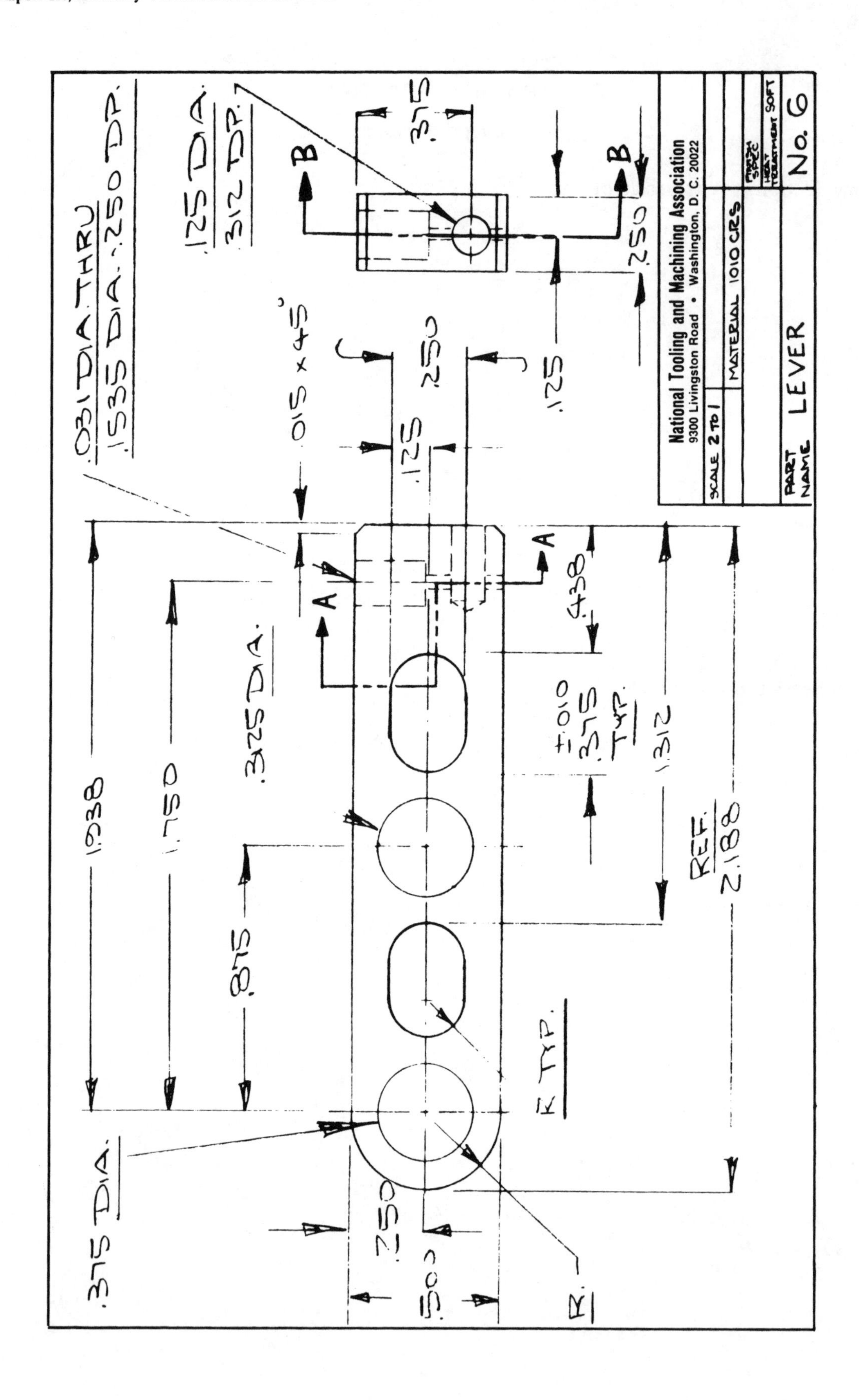
.031 DIA THRU
.1535 DIA. .250 DP.
.125 DIA.
.312 DP.
.375
B
.250
.015 x 45°
.250
.125
.125
1.938
1.750
.875
A
.438
±.010
.375
TYP.
1.312
REF.
2.188
R TYP.
.375 DIA.
.750
.500
National Tooling and Machining Association
9300 Livingston Road • Washington, D. C. 20022
SCALE 2 TO 1
MATERIAL 1010 CRS
FINISH SPEC
HEAT TREATMENT SOFT
PART NAME LEVER
No. 6

Chapter IV

DIMENSIONING ON DRAWINGS

Detail drawings contain all the dimensions necessary to make an object or part. Ideally, the drawing should be so accurate, and the dimensions so precise, that it can be interpreted one—and only one—way. Your understanding of the various types of dimensions and the methods used to represent them is essential in reading blueprints.

A *dimension* is a numerical value expressed in appropriate units of measure (fractional inches, decimal inches, millimeters, degrees, etc.) indicated on an engineering drawing along with lines, symbols and notes to define the dimensions and geometrical shape of an object or part. Terms associated with dimensions that you should understand are: basic dimension; reference dimension; nominal, basic, and actual sizes; limits; allowance; and tolerance.

A *basic dimension* (also known as a *mean dimension*) is a theoretical value used to describe the exact size, shape, or location of a feature (hole, face, edge, etc.). A *reference dimension* is a dimension used only for information (reference) purposes. *Nominal size* is the designation used for general identification. The *basic size* is the size from which the maximum and minimum sizes are determined by applying allowances and tolerances. For example, a tube may be called out as a ½ in. dia with a ± 0.010 in. variation. The *actual size* is the measured size; the maximum and minimum sizes become the *limits*. An *allowance* is an intentional measured difference between the maximum limits of mating parts. A *tolerance* is the total amount that a specific dimension may vary between its limits.

DIMENSIONING PLANS

Two plans of dimensioning are used on engineering drawings. One is called in-line (or point-to-point or chain) dimensioning; the other is called base-line (or datum dimensioning or the reference-line method).

In-Line Dimensioning

When the in-line dimensioning plan is used, each length is dimensioned, in sequence, from the end of the preceding one, as illustrated in Fig. 4-1. These dimensions are usually taken directly from the designer's sketch. However, this method allows a wide margin of error, because the tolerances tend to accumulate. Instead of averaging out, they usually add or subtract so that intermediate clearances may be disturbed.

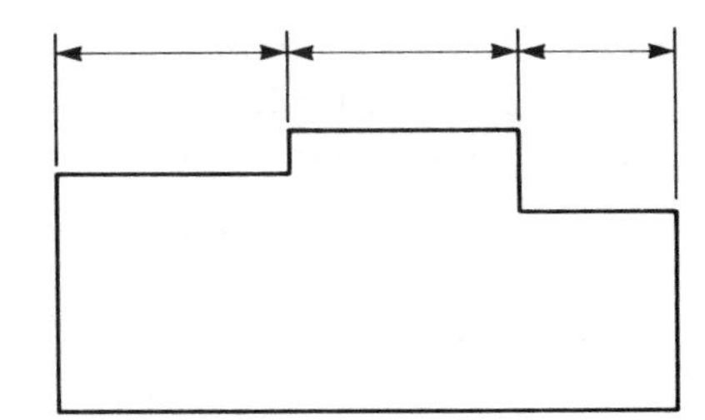

Figure 4-1 In-line dimensioning.

Base-Line Dimensioning

In the base-line dimensioning plan, all dimension lines extend in each direction from a datum, or base or reference line or plane (see Fig. 4-2),

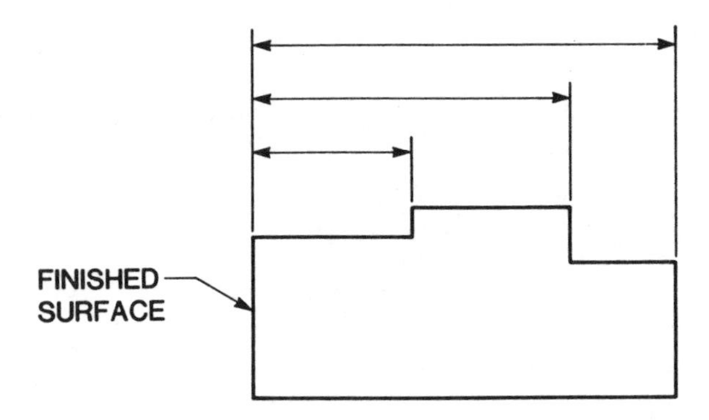

Figure 4-2 Base-line dimensioning.

which is usually the first machined surface. This method of dimensioning is extremely accurate—much more so than in-line dimensioning, because tolerances do not accumulate and a mating part will be within the specified limits.

Dimensioning Systems

The common dimensioning practice in the United States is to use inches and common fractions, but where more accuracy is required, decimal fractions are used. The use of decimal fractions in dimensioning allows for tolerances to thousandths of an inch (0.001 in), or to ten-thousandths of an inch (0.0001 in.).

The United States is in the process of converting from the English measurement system (inches, feet, yards, etc.) to the metric measurement system (millimeters, centimeters, meters, etc.). In much the same way as our system of paper money, metric measure is based on multiples of 10. You will learn more about metrics in Chapter VI.

The complete conversion of U.S. dimensions to the metric system will take many years. Meanwhile, you will often work with drawings that are dimensioned in *both* English and metric units. This method is called *dual dimensioning*. When dual-dimensioned, both the English and metric dimensions are given on the same drawing as shown in Fig. 4-3. Recently the conventional practice of writing these two dimensional systems together on drawings has given way to listing metric dimensions in a separate table (or conversion chart) keyed to the inch equivalents. Chapter VI covers this topic in detail.

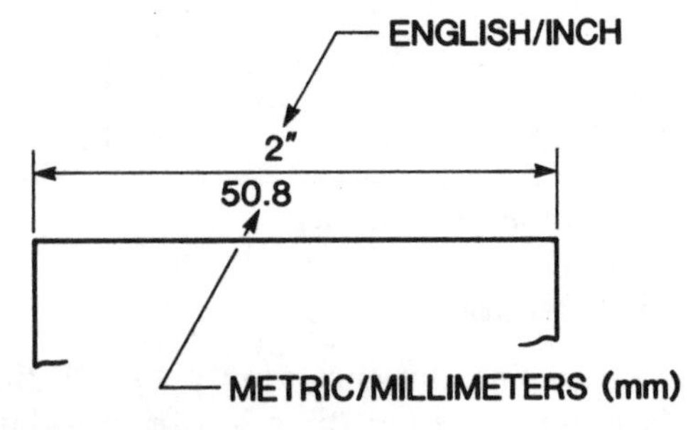

Figure 4-3 Dual dimensioning.

Drawing to Scale

The scale of a print refers to the size of the object on the drawing paper compared to its real-life size. The scale is the ratio between the size shown on the drawing and the actual size of the object. As you learned in Chapter I, an object may be drawn actual size or full scale if it will fit on the paper. If it is too large for the paper and is dimensioned in the English system, it may be drawn 1/2, 1/4, 1/8 actual size, etc. If the part is very small, it may be drawn 1½, 2, or 3 times larger than its actual size to make the detail easier to see and represent. The scale of a print is noted on the drawing itself (usually in the title block). If it is full scale, the note will read one of several ways: 1 = 1, 12″ = 1′, 1:1, or "Scale: Full," or simply "Full." Reduced or enlarged scale drawings will be noted ½ = 1 (meaning one-half actual size), 2 = 1 (meaning two times actual size), 3 = 1 (meaning three times actual size), etc.

However, drawings dimensioned in the metric system are scaled differently. If the object cannot be drawn full size on the paper, it may be drawn 1/2.5, 1/5, 1/10, 1/20, 1/50, 1/100, 1/200, 1/500, or 1/1000 actual size. If the object is small, it may be enlarged 2, 5, or 10 times its actual size.

BASIC DIMENSIONS

The general groupings of dimensions are: overall, size or detail, and position or location.

Overall dimensions give the entire length, height, and width of an object. They are the total of all included smaller dimensions, and are usually located on the outside of detail and position dimensions.

Size or detail dimensions give the sizes of diameters, widths, lengths, or heights of the part or object.

Position or location dimensions specify the location of the features of an object with respect to each other, such as the distance between centers, between a surface and a center, or between two surfaces.

Draftsmen usually follow certain rules or conventions when dimensioning drawings, although some objects have shapes, sizes, or complex portions that may require deviating from these rules. In general:

(1) Dimensions that apply to related views are placed between them.

(2) Dimensions are never placed directly on an object.

(3) Dimensions that apply to more than one view are placed on the view that most clearly illustrates the feature being dimensioned.

(4) Hidden edges are not used for dimensioning.

(5) All dimensions required to make the object are given without unnecessary duplication.

(6) Small dimensions are placed near the object, and larger ones are placed farther away.

(7) The symbol or abbreviation for inches is omitted from dimensions, and you can assume that the measurements are in inches unless stated otherwise.

(8) Numerals and letters are positioned so that you read them from the bottom of the drawing.

SPECIAL DIMENSIONS

In addition to the basic dimensions, drawings are dimensioned for the special features of an object or part, such as diameters in profile, radii, limits, tolerances, allowances, holes and threads, tapers and keyways.

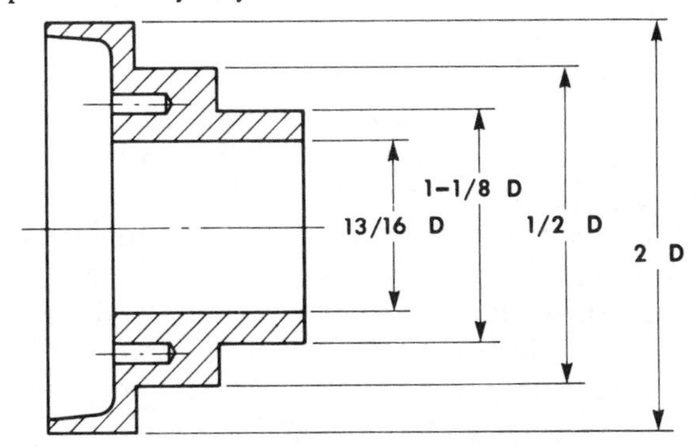

Figure 4-4 Dimensioning diameters in profile.

Diameters in Profile

Figure 4-4 shows the method used to dimension diameters shown in profile on a drawing. When diameters are shown as circles, one of two different dimensioning methods are used. Figure 4-5 shows the method used for smaller diameters, and Fig. 4-6 shows how larger diameters are dimensioned. Centerlines are used in all methods.

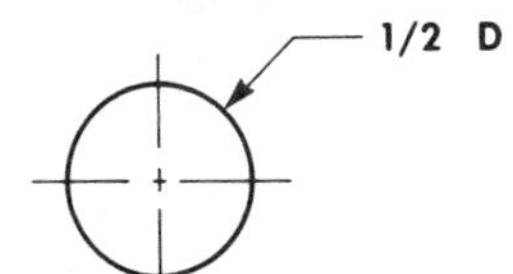

Figure 4-5 Dimensioning small circle.

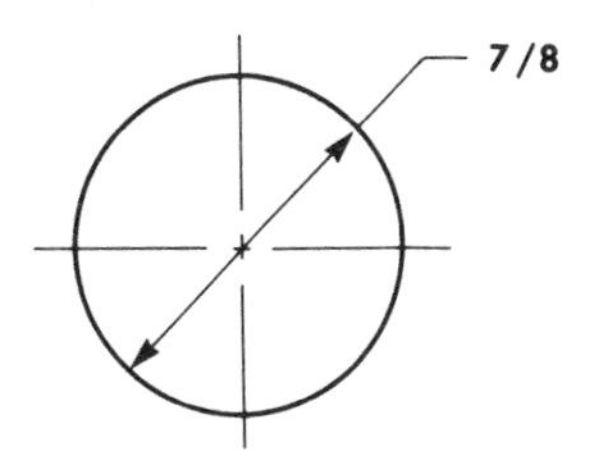

Figure 4-6 Dimensioning large circle.

Radii

Radii are dimensioned by drawing a radial dimension line to the center point of the radius as shown in Fig. 4-7. The letter "R" always follows the dimension of the radius.

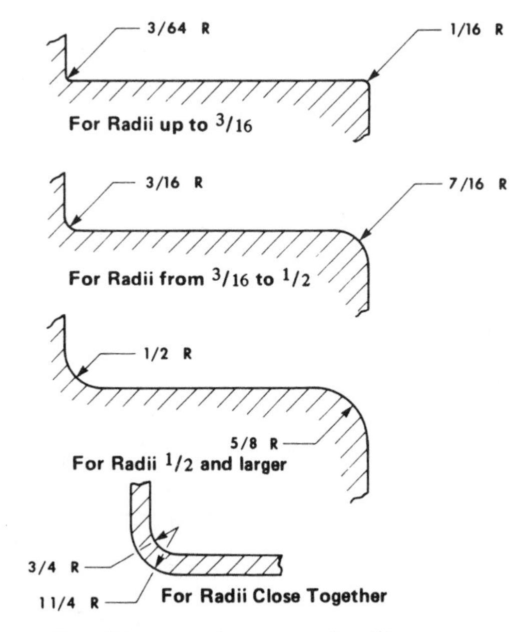

Figure 4-7 Dimensioning arcs and radii.

Angles and Chamfers

Figures 4-8 and 4-9 show how angles and chamfers are dimensioned on drawings. The dimension of the angle is placed inside the extension lines when space permits. Figure 4-9 illustrates the recommended dimensioning of chamfers, and the optional method used for 45 deg chamfers.

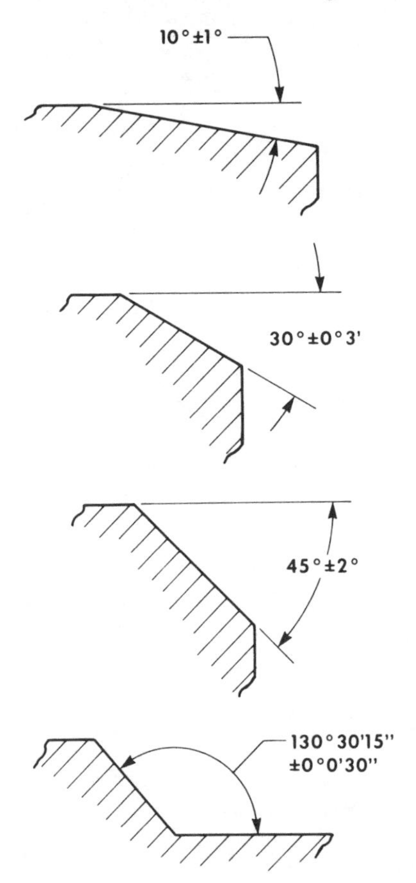

Figure 4-8 Dimensioning angles.

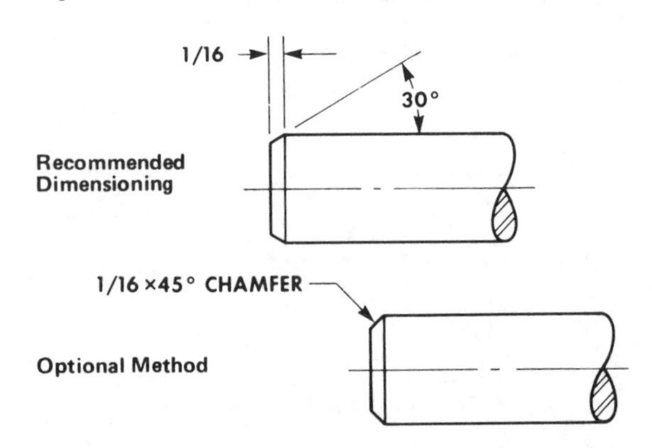

Figure 4-9 Methods of dimensioning chamfers.

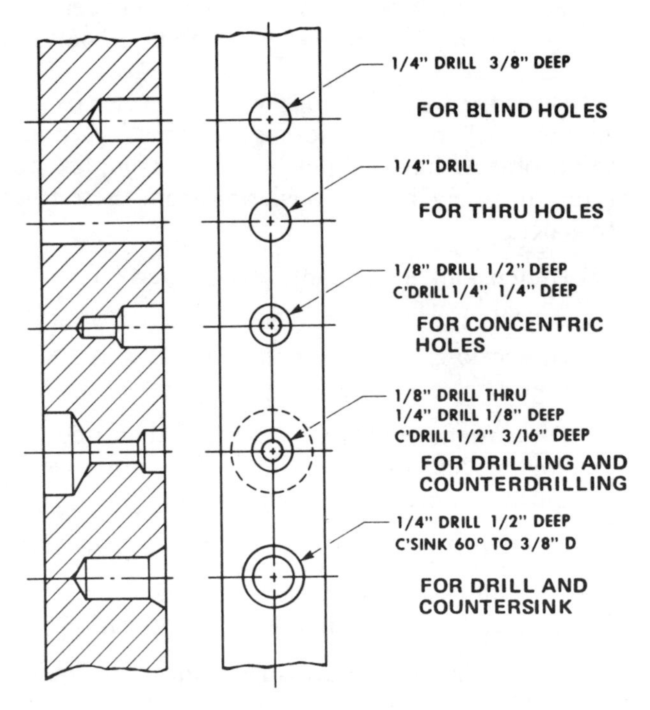

Figure 4-10 Dimensioning holes.

Limits, Tolerances, and Allowances

To make parts interchangeable, and to eliminate working to unnecessarily close measurements that are costly, the designer determines acceptable variations in size which he notes on the drawing. These "allowable errors" in size are known as tolerances, limits, and allowances. To interpret a drawing correctly, you must understand what each of these terms means, and how they are indicated on drawings. You will learn about tolerances, limits, and allowances in Chapter V.

Holes and Threads

As a machinist, you will be required to drill, countersink, counterbore, and tap holes, as well as to machine screw threads. To perform these operations properly, you must be able to read a drawing and recognize these operations by how they are illustrated (see Fig. 4-10). As general guidelines, the depth of blind holes is given on drawings, but does not include the drill point. Countersunk holes will have the included angle and diameter of the countersink given. The abbreviations "C'Drill" (counterdrill), "C'Bore" (counterbore), and

"C'Sink" (countersink) are written along with the size of the hole.

Threaded holes and shafts are depicted in different ways on drawings, as was illustrated in Figs. 2-2 and 2-3. Figure 4-11 shows a typical screw thread, and gives the names of its various parts. Note that the grooved part of the thread is called the *root*. The raised part is the *crest*. The distance from one crest to the next is called the *pitch* of the thread. The *thread angle* is the included angle between the sides of the thread, measured as shown.

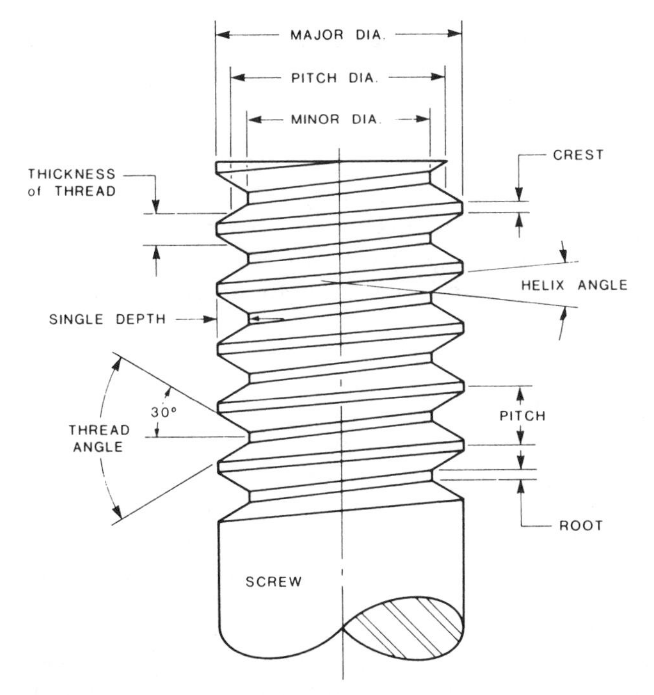

Figure 4-11 Typical screw thread.

Thread Notation

A very important specification that you will find on blueprints is thread notation. It is a type of formal note that provides all the necessary information to describe a thread. Screw thread specifications are written, in sequence: nominal size, number of threads per inch, thread series symbol, thread class symbol, and hand.

Nominal size is the major diameter, and is specified as the fractional diameter, screw number, or their decimal equivalents. Below ¼ in., thread diameters are identified by numbers or their decimal equivalent. Above ¼ in., thread diameters are expressed in inches and fractions of inches.

Number of threads per inch is the pitch specified in inches (sometimes abbreviated "TPI").

Thread series are groups of diameter/pitch combinations identified by the number of threads applied to each diameter in the series (see the next section, "Unified Screw Thread Standards").

Thread Classes are grouped by the amounts of tolerance and allowance (see the next section, "Unified Screw Thread Standards").

Hand refers to right-hand and left-hand threads. A right-hand thread advances into the material when turned clockwise. A left-hand thread advances into the material when turned counterclockwise. If a thread is not specified on a drawing as a left-hand thread (LH), you can assume that it is a right-hand thread.

Unified Screw Thread Standards

The American National form (formerly known as the United States Standard) was used for many years for most screws, bolts, and other threaded products manufactured in the United States. It was replaced by the American National Standard for Unified Screw Threads in 1949, and revised in 1974. The new unified screw-thread standards published by ANSI (the American National Standards Institute) as *American Standard Unified and American Screw Thread Publication B1.1–1974* covers the thread series and thread classes used in screw thread specifications today. It includes six Standard thread series and three Special thread series.

Series. Each series specifies a particular number of threads per inch for a particular diameter.

Coarse-Thread Series (UNC and NC)—for general use, especially where rapid assembly is required, and for gray iron, soft metals, and plastics.

Fine-Thread Series (UNF and NF)—for applications requiring greater strength or where the length of engagement is limited.

Extra-Fine-Thread Series (UNEF and NEF)—for highly stressed parts, and where internal threads are required in thin-walled fasteners.

8 Thread Series (8N)—substitutes for the Coarse-Thread Series for diameters larger than 1.0 in.

12 Thread Series (12 UN and 12N)—a continuation of the fine-thread Series for diameters larger than 1½ in.

16 Thread Series (16 UN and 16N)—a continuation of the Extra-fine-Thread Series for diameters larger than 2 in.

8 UN, UNS, and NS (the three Special series)—cover nonstandard combinations of diameter and pitch, and are used only in cases where it is impossible to use the Standard series.

The threads common to American, British, and Canadian standards are known as Unified, and carry the letter "U" in the thread series symbol. Those without the "U" are American standard only.

Classes. The eight thread classes of tolerance or fit in the ANSI standard distinguish between external and internal threads. Classes 1A, 2A, and 3A apply to external threads only. Classes 1B, 2B, and 3B apply to internal threads only. Classes 2 and 3, which are used with American Standard threads only, apply to both external and internal threads.

Classes 1A and 1B—provide liberal allowance for ease of assembly, even when threads are dirty or slightly damaged.

Classes 2A and 2B—for production of bolts, screws, nuts, and other commercial fasteners; permit external threads to be plated.

Classes 3A and 3B—for close-tolerance work where no allowance is required.

Classes 2 and 3—retained as American standard pending industry transition to the Unified classes.

A typical specification is written "¼-20 UNC-3A" and means ¼ in. nominal diameter, pitch of 20 threads per inch, Unified Coarse Series, Class 3 fit external right-hand thread. Other typical specifications and how they are noted on drawings are shown in Figs. 4-12 and 4-13.

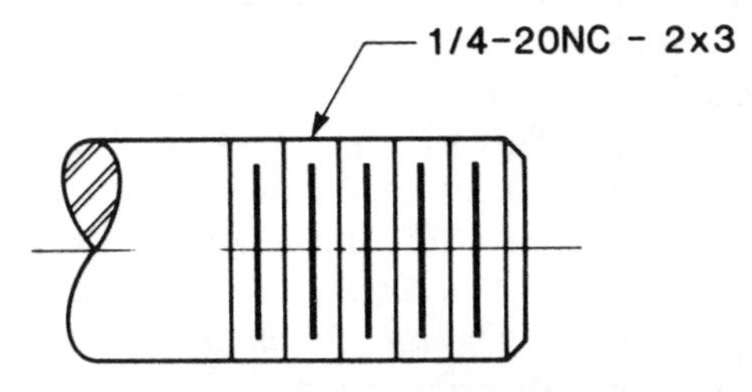

Figure 4-12 External threads dimensioned with thread notation meaning: ¼ in. Nominal dia, 20 TPI, American Coarse Series, Class 2 fit, 3 in. long.

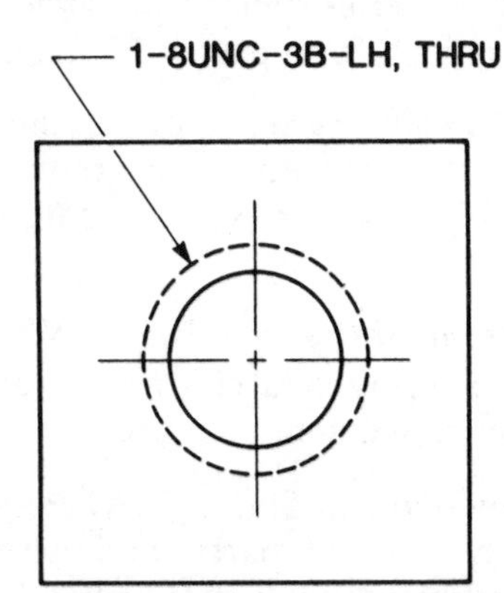

Figure 4-13 Internal threads dimensioned with thread notation meaning: 1.0 in. dia, 8 TPI, Unified Coarse Series, Class 3 fit, interior left-hand thread, completely through part.

Tapers and Keyways

The dimensions of tapers are given in taper per foot, measured on the axis rather than on the slope of the taper, or taper per inch. *Taper per foot* is the difference in diameter, in inches, for 1 ft of length. Three basic dimensioning methods are used:

Standard tapers are dimensioned by one diameter or width, the length, and a note on the drawing designating the taper by American Standard taper number.

Special tapers, whose slope is specified in taper per foot, are dimensioned by one diameter and the length, or the diameters at both ends of the taper are given and the length is omitted.

Precision tapers are dimensioned differently. The taper surface (either external or internal) is specified by a diameter given at a particular distance from a surface and the slope of the taper.

How keyseats and keyways are dimensioned depends upon the drawing's purpose. For unit production, nominal dimensions are usually given, with fitting to be done by the machinist. For quantity production, the limits of width and depth are usually given.

SUMMARY

Size and position of objects or features on drawings are dimensioned two ways, by in-line and base-line plans. Base-line dimensioning is preferred, because it does not accumulate error like in-line does.

Dual dimensioning is the practice of including a metric equivalent with its English measure counterpart within the same dimension lines.

Threads are designated on drawings with a special code or notation. All necessary information needed to make a thread is contained in this notation, i.e., major diameter, threads per inch, type of thread, etc.

Hole sizes, spacing, tapers, and hole features are all noted on a drawing in a prescribed manner. Examples of these and their meaning are included in this chapter.

DIMENSIONING PRACTICES

The following examples show the most common dimensioning techniques/practices for special features, such as holes, chamfers, tapers, threads, etc. With each example is a written interpretation of the dimensions given on the drawing.

Holes, Chamfers, and Angles

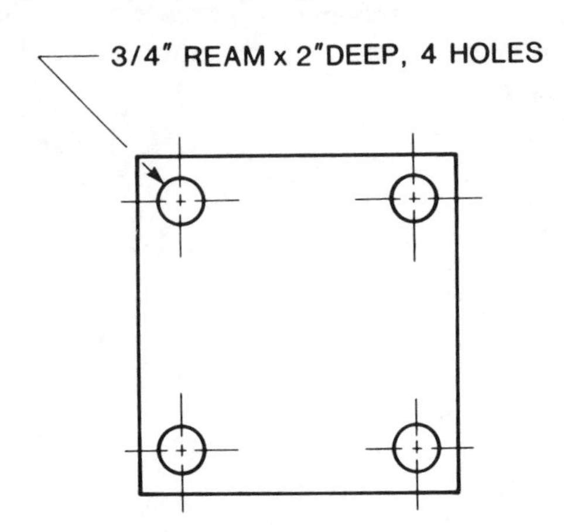

Drill and ream four holes 3/4″ in diameter and 2″ deep.

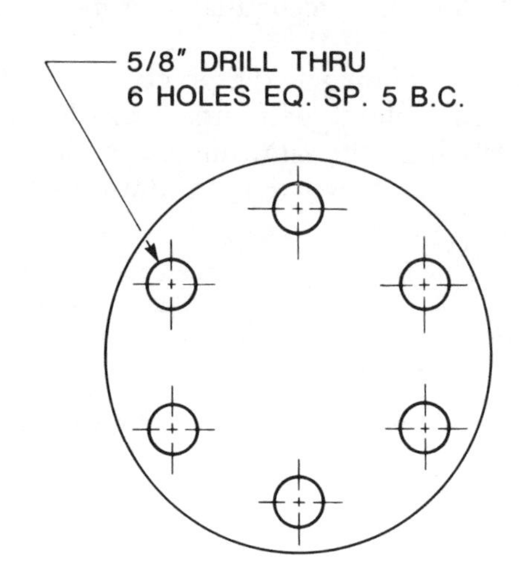

Drill six holes 5/8″ diameter, equally spaced on a 5″ bolt circle.

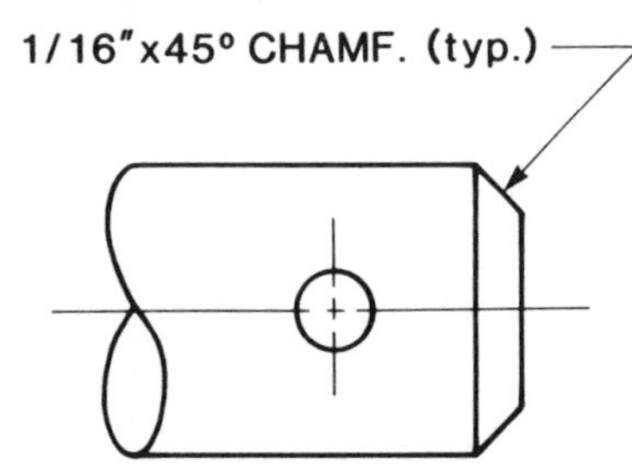

45 deg chamfer, 1/16″ long; chamfer is typical unless otherwise noted.

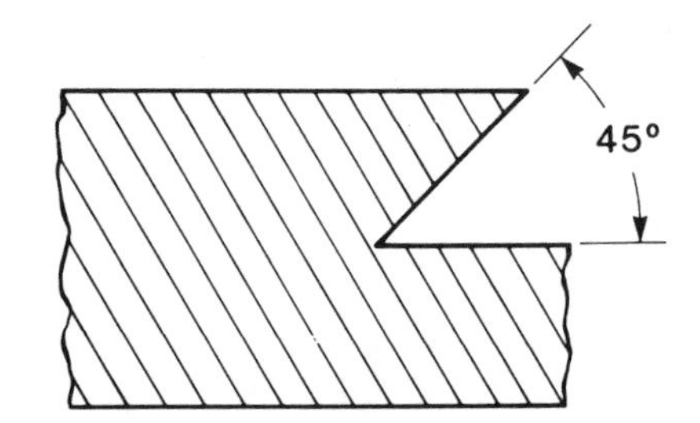

Cut an angle within extension lines of 45 deg.

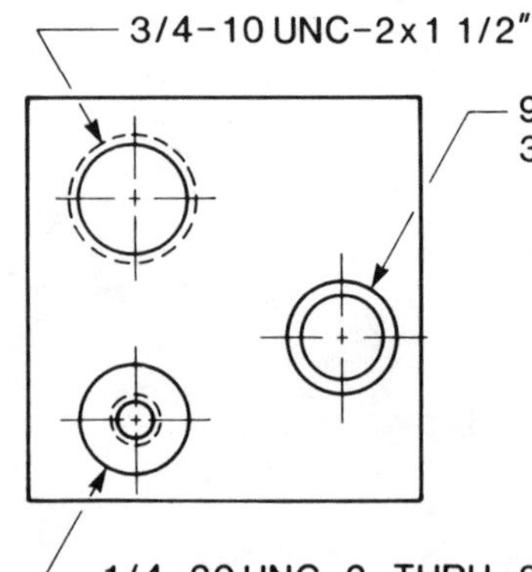

(1) 3/4″ diameter – 10 TPI tapped holes, 1 1/2″ deep.
(2) 9/16″ diameter drill through 3/4″ counterbore, 5/8″ deep.
(3) 1/4″ diameter – 20 TPI tapped holes through part, 82 deg countersink.

Schematic and Simplified Hole Presentations

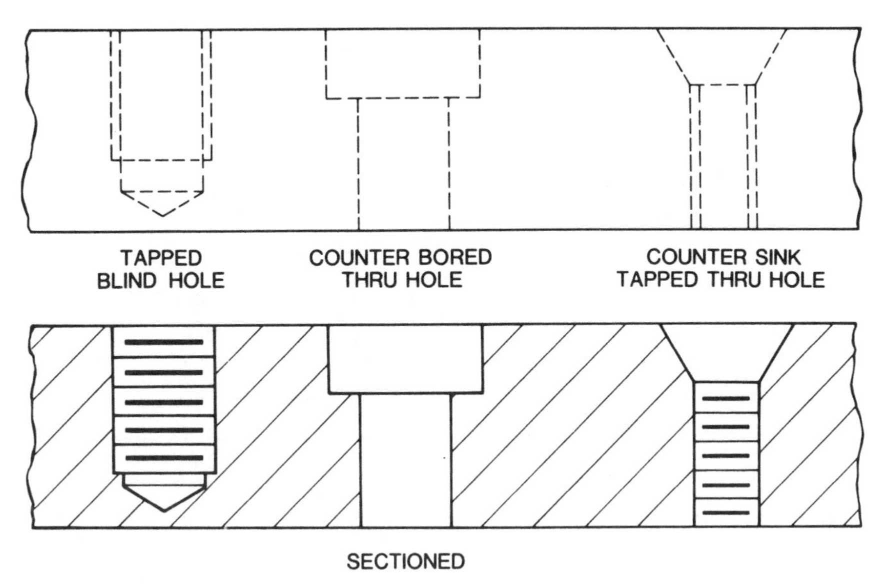

1 1/2" DIA. THRU, 20 DIA.
S.F.x1 16" DEEP, 2 HOLES

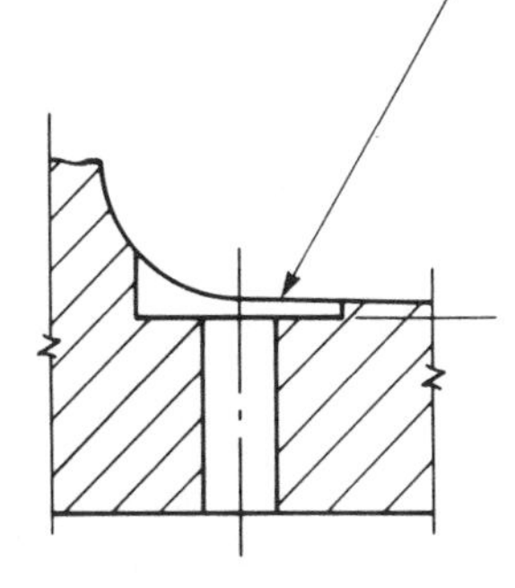

1½" diameter drill through part, 2" diameter spot face 1/16" deep, in 2 places.

Tapers and Keyways

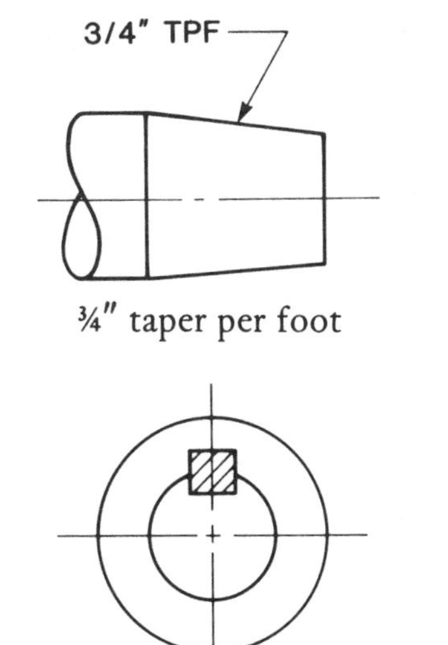

¾" taper per foot

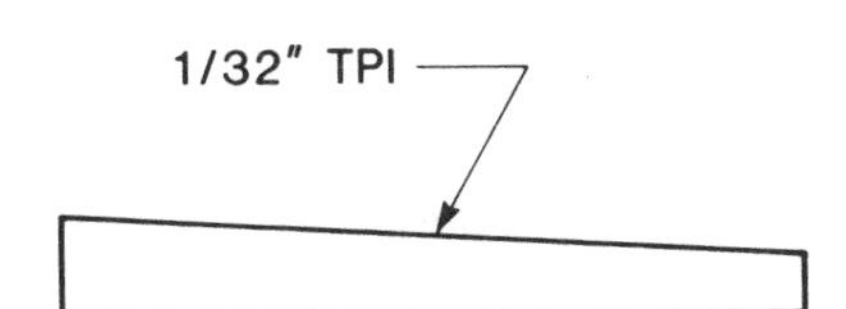

1/32" taper per inch

3/16" wide x 3/32" deep keyway

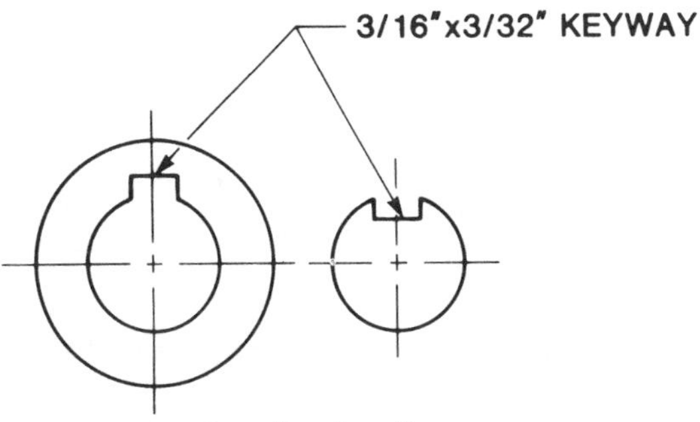

3/16" x 3/32" keyway

Diameters and Radii

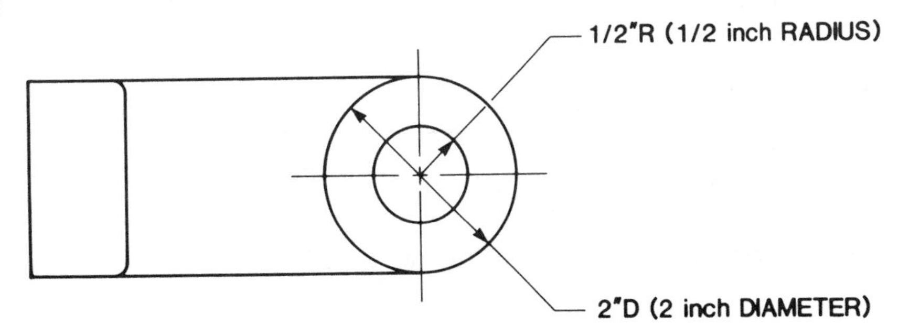

TRAINING PRACTICE

Answer each of the following questions completely. Keep your answers brief. Refer to the blueprints associated with each set of questions.

Training Exercise #1

Blueprint No. 1

1. Name the three views in this drawing.
   a. ____________________
   b. ____________________
   c. ____________________

2. How many holes are shown? ____________________

3. What characteristic do these holes have in common? ____________________
   ____

4. Which dimensioning plan does this print use? ____________________
   ____

5. Give the thickness, width, and length of this part. ____________________
   ____

6. How deep are the holes drilled? ____________________
   What is this kind of hole called? ____________________

7. Starting from the left, what is the distance between hole centers? ____________________
   ____

8. List the steps in laying out this part. ____________________
   ____
   ____
   ____
   ____

9. Why do you think View III does not show hidden lines? ____________________
   ____

10. What is the part made from? ______________________________

11. How many parts are required? ______________________________

12. What does the note direct you to do? ______________________________

____

13. What is the line called that passes through View III? ______________________________

____

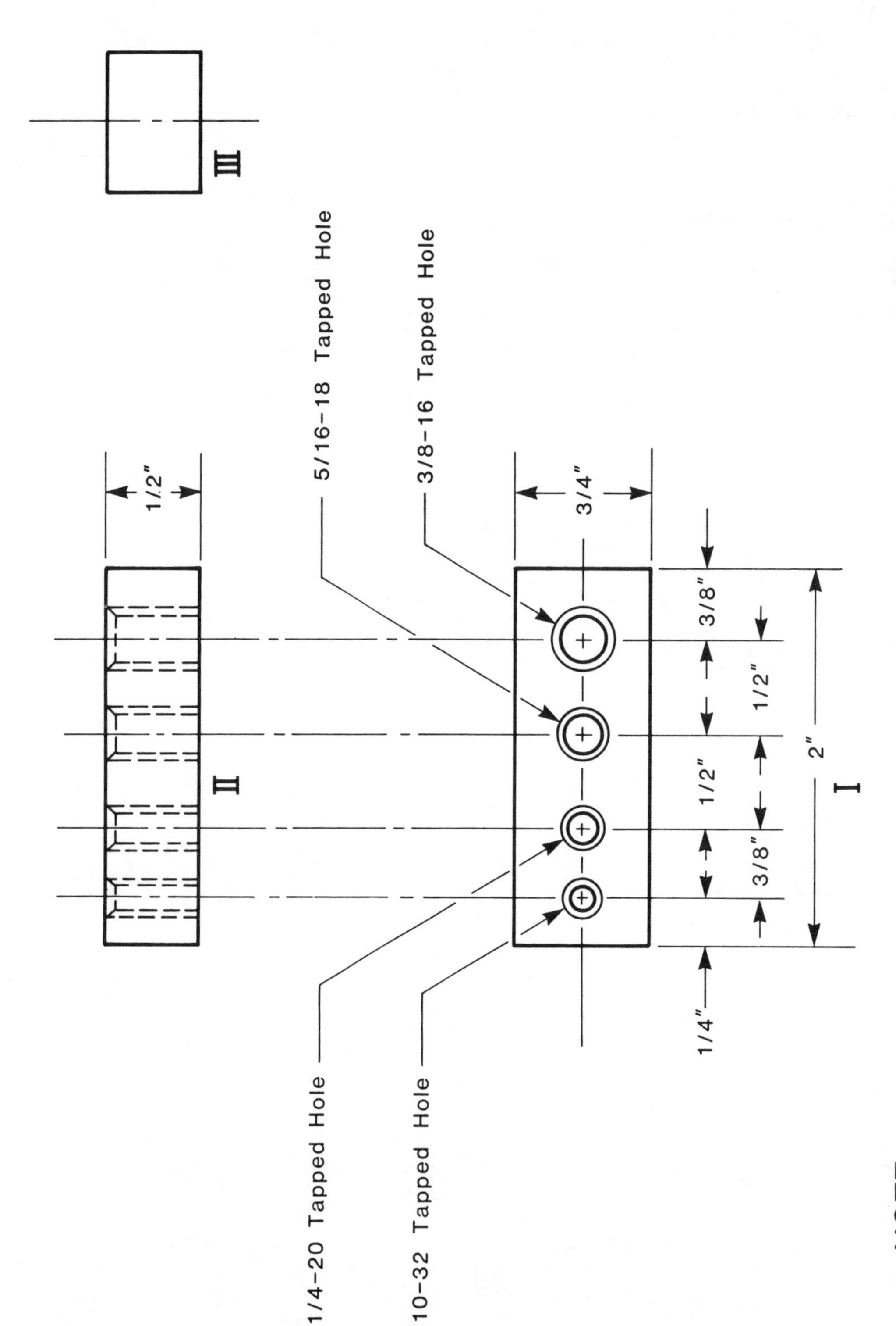

1/2"
III
II
I
5/16-18 Tapped Hole
3/8-16 Tapped Hole
3/4"
3/8"
1/2"
2"
1/2"
3/8"
1/4"
1/4-20 Tapped Hole
10-32 Tapped Hole
NOTE:
Countersink Holes From One Side
CLAMP BLOCK
3 Req. Tool Steel SAE 1045
Print Number 1

Training Exercise #2

Blueprint No. 2

1. What are the names of the views presented in the drawing? ______

2. Why are there only two views? ______

3. Which dimensioning plan is used? ______

4. What does the letter "G" mean on this print? ______

   And where would you expect to find such information? ______

5. How much stock is allowed *on the diameter* for this finishing operation? ______

6. What is the machinist directed to do to the 1 1/8 dia? ______

7. What is the diameter of the groove? ______

8. What do the hidden lines in the front view represent? ______

9. What do the hidden lines in the end view represent? ______

10. How large or how small can the .562 diameter be and still be acceptable? ______

11. What operation must be performed on the smallest diameter? ______

12. What material is this part made from? ______

13. How many 1/16 x 45° chamfers are required, and where are they located? ______

14. What is the smallest dimension that the overall length of the part can measure? ________________

____

15. Counting from left to right (from the small end of the part), what does the second "G" indicate?

____

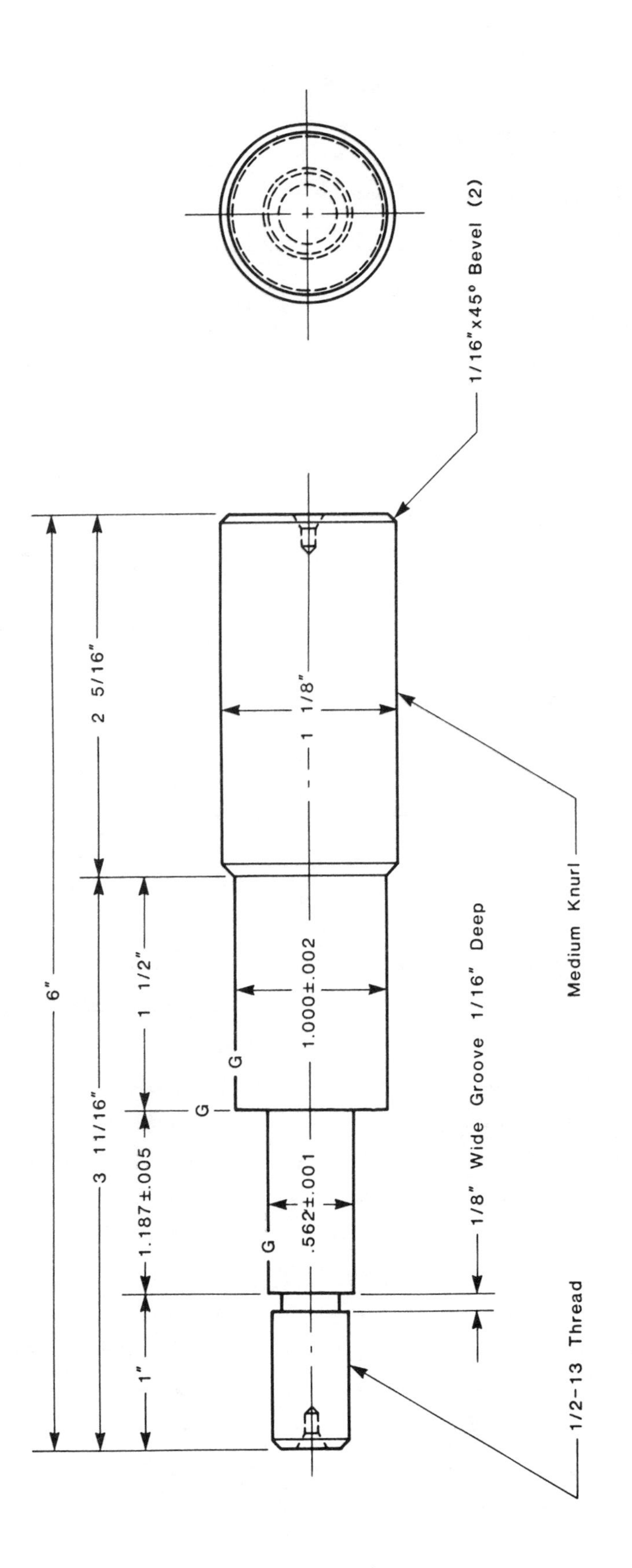

LATHE and CYLINDRICAL GRINDER

Cold-Rolled Steel-Soft
Print Number 2

NOTE:

Allow .010 on a Side for Grinding on Surfaces Specified G

Tolerances:
Fractional ±1/64"

Training Exercise #3

Blueprint No. 3

1. Give the name of each view shown.
   I. ______
   II. ______
   III. ______
   IV. ______

2. What is:
   a. The print name ______
   b. The print number ______
   c. The material the part is made from ______

3. At what angle is the slot at the top of the part? ______

4. How far off-center is the tool bit slot? ______

5. How deep should the student identification pocket be cut? ______

6. What is the dimension of the largest diameter? ______

7. If this drawing were base-line dimensioned, which dimension should be eliminated in View IV: (I) or (H)? ______

8. What does (J) in View III mean? ______

9. What are four views of the object shown? ______

10. What kind of line is Line:
    (E) ______
    (F) ______
    (G) ______
    (H) ______
    (K) ______
    (L) ______

11. What is the scale of the drawing? ____________________

12. If the scale was 2 : 1 (or double), how long would the 2½ dimension be on the drawing? ____________________

13. In the 1/16 x 45° chamfer in View IV, what does the 1/16 dimension refer to? ____________________

14. Surface (A) in the top view represents which surface in View II? ____________________

15. Surface (B) in View IV represents which surface in View III? ____________________

16. What is the dimension of (C) in View II? ____________________

17. Why is the .750 $^{+.000}_{-.001}$ dimension in View III a decimal fraction and not a common fraction? ____________________

18. How deep is the tool slot? ____________________

19. Line (D) in View I represents which line in View IV? ____________________

20. Which view did you select as the front view, and why? ____________________

21. What indicates that the .750 dia in View III is concentric with the 1½ dia in View I? ____________________

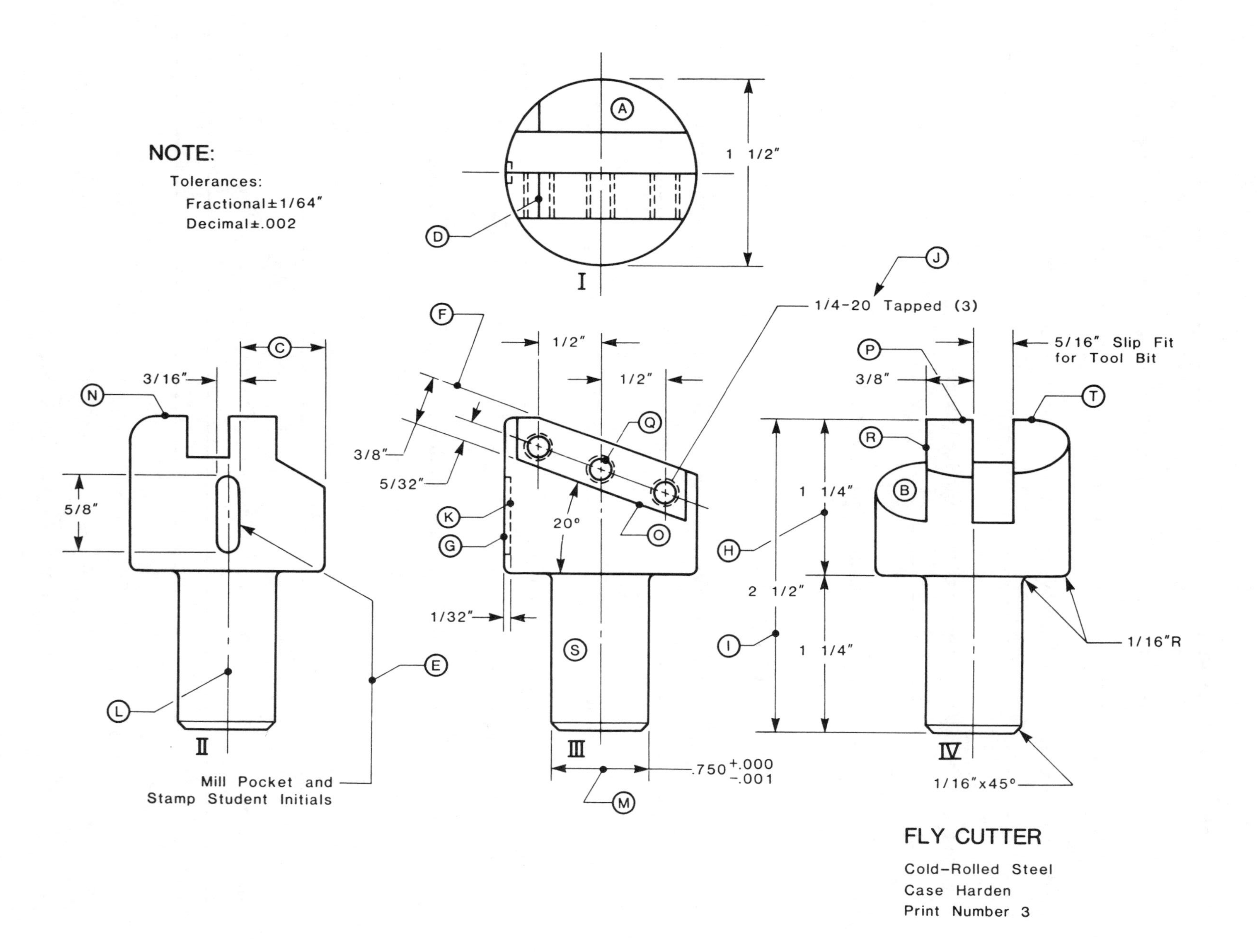
NOTE:
Tolerances:
Fractional ±1/64"
Decimal ±.002
1 1/2"
I
3/16"
5/8"
Mill Pocket and
Stamp Student Initials
II
1/2"
1/2"
3/8"
5/32"
20°
1/32"
III
.750 +.000 -.001
1/4-20 Tapped (3)
5/16" Slip Fit
for Tool Bit
3/8"
1 1/4"
2 1/2"
1 1/4"
1/16"R
IV
1/16"x45°
FLY CUTTER
Cold-Rolled Steel
Case Harden
Print Number 3

Chapter V

GEOMETRIC DIMENSIONING AND TOLERANCING AND SURFACE TEXTURE REQUIREMENTS

Mass production has brought about the need for interchangeable parts in manufactured products. Mating or adjacent parts that are manufactured and assembled in entirely different factories must fit together and function perfectly. *Precision* is the term used to describe the degree of machining accuracy required to ensure the functioning of mating parts as intended.

So that parts can be interchanged, dimensions given on drawings are specified in a special way that enables machine operators in widely-separated shops to produce accurate and interchangeable parts according to the same design. In this special way of dimensioning parts on drawings—called *tolerancing*—the basic dimensions are adjusted slightly to accommodate the inevitable mechanical, measurement, and human error inherent in manufacturing any part or product. These adjustments—called *tolerances*—specify the allowable amount of error (between decimal limits) that can exist and still permit the parts to function satisfactorily. The allowable amount of error is called tolerance.

TOLERANCING

Tolerance is defined as the total amount of allowable variance (error) from a perfect measurement, and is the difference between the maximum and minimum limits imposed on a basic dimension. The tolerance on any given dimension varies according to the degree of accuracy or precision necessary for the particular part surface or feature. In precision machining, tolerances are expressed in decimal numbers, usually to at least three places.

Engineering tolerances are divided into six groups: size tolerances, form tolerances, orientation tolerances, profile tolerances, runout tolerances, and location tolerances.

Size tolerances specify limits of size on dimensions such as length or height, thickness or depth, diameter, and angle (see Chapter IV), in limit dimensioning and plus and minus tolerancing forms.

Form tolerances are used to control certain geometric characteristics of a part feature, such as flatness, straightness, roundness or cylindricity. They are specified in notes or by symbols.

Orientation tolerances are used to specify *relationships* of features. One feature is selected as a datum feature, and the orientation tolerance relates to it. Perpendicularity, angularity and parallelism are orientation tolerances.

Profile tolerances are used to specify a permissible deviation from the desired outline; usually an irregular shape where other geometric controls are inappropriate.

Runout tolerances indicate the permissible error of the controlled feature surface when rotated about a datum axis. There are two types of runout control: *circular* runout and *total* runout.

Location tolerances control the relative positions of mating or adjacent features to obtain interchangeability. They are specified on dimensions in limit dimensioning and plus and minus tolerancing forms, or by a combination of symbols.

Prior to World War II, the only system of tolerancing used in the United States was confined to adding numerical values (in a prescribed method of notation) to the nominal sizes and locating dimensions on a drawing, and indicating form tolerances in notes connected by leaders. The difference between mating parts, or allowance, was indicated by a standardized classification of fits. This sytem is still in use today, and is referred to as *general tolerancing.* However, another system of tolerancing—called *geometric dimensioning*—was developed by the British and U. S. military during the war, and a refinement of this sytem is now an American National Standard.

To understand and interpret tolerances properly, you must understand certain associated terms:

Dimension is a numerical value expressed in appropriate units of measure, and indicated on a drawing along with lines, symbols, and notes to define the geometric size, shape, and location of an object or a feature.

Nominal size is the approximate size used for the purpose of general identification.

Basic dimension is a numerical value used to describe the theoretically exact size, shape, or location of a feature. It is the basis on which permissible variations are established by means of tolerances, notes, or feature control frames. A basic dimension is indicated on drawing by enclosing it in a box [4.465], or the notations BSC or BASIC.

True position is the theoretically exact location of a feature established by basic dimensions.

Basic size is the theoretical size from which the maximum and minimum limits of size are determined by the application of allowances and tolerances.

Limits of size is the name given to the extreme maximum and minimum sizes specified by a toleranced dimension.

Actual size is the measured size of a feature.

Virtual condition (sometimes called *virtual size)* is the boundary generated by the collective effects of the Maximum Material Condition (MMC) limit of a feature and any applicable geometrical tolerance.

Datum is the name given to a point, line, plane, cylinder, or other geometric shape that is taken to be exact for purposes of computation, and from which the location or geometric relationship (form) of part features is established.

Maximum material condition is a modifier that specifies the maximum limit of the size of an external feature, or the minimum limit of the size of an internal feature: for example: minimum hole diameter or maximum shaft diameter. Abbreviated MMC, it is the most critical specified interchangeable size of a part feature. An MMC feature is derived from its toleranced dimensions so that it contains the maximum amount of material with respect to any portion of the feature that will affect a geometrical tolerance.

Least material condition is a modifier applied to parts that require precise positioning so that they will assemble properly. Abbreviated LMC, an LMC feature is derived from its toleranced dimensions so that it contains the least amount of material with respect to any portion of the feature that will affect a position or form tolerance. For a hole, LMC represents the high dimensional limit. For a shaft outside diameter, LMC represents the low dimensional limit.

Regardless of feature size is a modifier that indicates that the tolerance cannot vary. Abbreviated RFS, it means that the geometrical tolerances of the feature are fixed and must not be exceeded, regardless of the finished size of the feature. This modifier is used, for example, on a drawing for a part designed to rotate and which must be balanced.

*Allowance** is the prescribed difference between the MMC limits of size of mating parts. If the mating parts have minimum clearance between them, the allowance is positive and is called a *clearance fit.* If the mating parts have maximum interference between them, the allowance is negative and is called an *interference fit.*

Fit is the term commonly used to describe the range of tightness or looseness that results from applying a specific combination of allowances and tolerances to the design of mating parts. There are

* The term "allowance" is also applied to excess stock deliberately left on a piece of material that will be further machined to the final size. For example, a part that is to be heat-treated is usually made oversize, and then ground to the final size following heat treating. The excess stock is called *grinding allowance.*

four basic kinds of fits: clearance fit, interference fit, transition fit, and line fit.

Clearance fit is a fit whose limits of size assure clearance between assembled mating parts—axially, radially, or both.

Interference fit is a fit whose limits of size result in a degree of tightness between two mating parts. In the case of a hole and shaft, the shaft will be larger than the hole, resulting in an actual interference of metal that will require a force or press fit.

Transition fit is a fit whose limits of size can lead to either clearance or interference. This means that a shaft may be either larger or smaller than the hole in a mating part.

Line fit is a fit whose limits of size are so specified that surface contact or clearance will result when mating parts are assembled.

Maximum material limit is the limit of size that provides the maximum amount of material for a part. It is usually the maximum limit of size of an external dimension, or the minimum limit of size of an internal dimension.

Minimum material limit is the limit of size that provides the minimum amount of material for a part. It is usually the minimum limit of size of an external dimension, or the maximum limit of size of an internal dimension.

Running and sliding fits are names for special types of fits applied to cylindrical parts whose surfaces are in contact, and which move on each other. If two surfaces are fitted to slide on each other without losing lateral motion, the fit is called a *sliding fit* or *slip fit*. (Examples are the cross and transverse slides of many machine tools.) If two surfaces are fitted with sufficient clearance to allow for lubrication and movement, the fit is called a *running fit*. (Examples are the running bearings of spindles, crankshafts, line shafts, etc.).

Location fit is another special type of fit applied to cylindrical parts. It is intended to determine only the location of the mating parts. It may provide rigid or accurate location (as with interference fits) or provide some freedom of location (as with clearance fits).

Force fit, press fit, or shrink fit are three names for a special type of interference fit applied to cylindrical parts that requires parts to be forced together under considerable pressure.

Basic hole and *basic shaft systems* are terms used to identify which mating part is taken as a standard size. In the basic hole system, the minimum size of the hole is taken as the base from which all variations are made. The hole can often be made with a standard tool. In the basic shaft system, the maximum shaft size is taken as the basic size. This system is used where several different fits of nominal size are required on one shaft; for example, when bearings are fitted to line shafting.

General Tolerancing

Detail and assembly drawings give both dimensions and tolerances. The tolerance may be general or specific. A *general tolerance* is a tolerance that applies to all dimensions on a drawing, unless otherwise noted. It is used wherever possible, and is written on the drawing in the title block or in a general note. A general tolerance may be expressed as a common fraction (such as ± 1/64), or as a decimal number (such as ± .005). *Specific tolerances* are attached to a particular dimension, as required by the degree of manufacturing accuracy.

Two main methods are used in general tolerancing to indicate tolerances of size and location on a drawing: limit dimensioning and plus and minus tolerancing.

Limit dimensioning is used to tolerance closely-fitting mating or adjoining parts to ensure interchangeability. Limits are always expressed in decimal numbers. The limits imposed on a dimension depend on the accuracy and clearance required for the parts to function satisfactorily.

For plain cylindrical (non-threaded) parts, limits are figured on the basic hole system (more common) or the basic shaft system. Two manufacturing limits are specified for the hole and two for the shaft, as shown in Fig. 5-1A. Standard reference handbooks or ANSI Standard B.4—1976, R 1974 contain tables that give the preferred limits and fits for plain cylindrical parts.

In limit dimensioning on a drawing, the upper limit (maximum value) is usually written over the lower limit (minimum value), as shown in Fig. 5-1A. Or, when written in a single line, the lower limit precedes the upper limit, separated by a short dash or hyphen.

Plus and Minus Tolerancing. Tolerances are expressed by specifying the nominal dimensions, followed by either a unilateral or bilateral tolerance. The nominal dimension may sometimes be a common fraction or a mixed number (whole number combined with a common fraction), but decimal numbers are always used in precision work.

Plus and minus tolerances may be expressed in one of two ways: the dimension precedes either a plus (+) expression of tolerance written over a minus (−) expression of tolerance, or the dimension precedes simply a (±) expression of tolerance.

Unilateral Tolerances. A unilateral tolerance allows variation in one direction only from a specified size, form, or location. When a piece can be made *only larger* or *only smaller,* it is said to have unilateral tolerance, as shown in Fig. 5-1B.

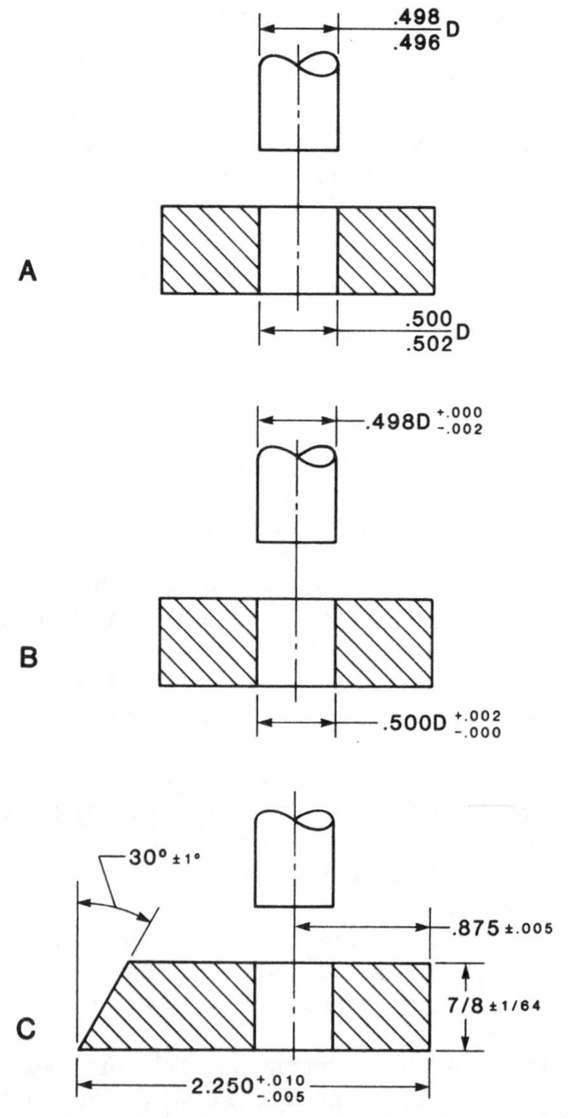

Figure 5-1 Methods of indicating tolerance in general tolerancing.

For example, if the size of a shaft is 1.000 in. and tolerance is expressed as 1.000 + .001 in. and −.000 in., this is unilateral tolerance.

Bilateral Tolerances. A bilateral tolerance permits variation in *both* directions from the specified size, form, or location. For example, when a piece can be made *either* larger or smaller, it is said to have bilateral tolerance. In a bilateral tolerance, the plus and minus values are usually equal, as shown in the expression ".875 ±.005" in Fig. 5-1C. The lower tolerance expression in Fig. 5-1C shows how bilateral tolerance is indicated when the plus and minus values are not equal.

In plus and minus tolerancing, you can find the maximum limit of size by adding the plus amount to the dimension, and you can determine the minimum limit of size by subtracting the minus amount from the dimension. You can determine total tolerance by adding the plus and minus values together.

Geometric Dimensioning and Tolerancing

The geometric dimensioning and tolerancing system is used:

(1) When part features are critical to function or interchangeability,

(2) When functional gaging techniques are to be used in inspection,

(3) When the use of datum references will help to ensure consistency between manufacturing and gaging operations, and/or

(4) When the standard interpretation or tolerance is not readily implied.

In this system, geometric tolerances specify how far part features can vary from the perfect geometry implied by the drawing.

A **form tolerance** states how far actual surfaces may vary from the desired geometric form of the controlled feature implied by the drawing. Form tolerances control the geometric characteristics of a feature of: straightness, flatness, roundness and cylindricity. Orientation tolerances control perpendicularity, angularity and parallelism. Runout tolerances control either circular or total runout. Profile tolerances control either profile of a surface or profile of a line. These characteristics may be specified independent of, or dependent upon, the size and other tolerances.

A **position tolerance** states how far individual features may vary from the perfect location implied by the drawing. Positional tolerances provide an effective means for controlling the relative location of mating features where interchangeability is a definite requirement. They control the geometric characteristics of position and concentricity. Positional tolerances serve to relate feature size to location, thereby ensuring interchangeability. They also include some form and orientation control.

Geometric tolerancing is the general term applied to form, orientation, runout, profile and positional tolerances. The dimensions used in this system are basic, rather than nominal, as in general tolerancing. Recall that basic dimensions describe theoretically exact sizes, locations, and relationships. Basic dimensions are, by definition, untoleranced dimensions that first locate from the *datum,* which may be a plane, a line, a point, or an axis. However, basic dimensions are not practical unless they are related to geometrical tolerances. The methods used to indicate tolerances in geometric tolerancing differ from those used in the general tolerancing system.

Sizes and positions are determined, and distance is measured, from a datum, which is indicated on the drawing by a symbol. Letters of the alphabet are used as *datum feature symbols* for reference. A datum feature symbol on a drawing consists of the datum reference letter contained in a box.

A positional tolerance specifies either the width or diameter of a tolerance zone within which an axis or centerplane of the controlled feature must lie for the final part to satisfy the accuracy requirements for proper functioning and interchangeability. The tolerance applies in all directions within that zone from *true position,* which is the theoretically exact location of a feature established by basic dimensions. Because such information cannot be conveyed clearly or compactly by dimensions and notes, it is indicated by standardized geometric characteristic symbols, as shown in Fig. 5-3. On a drawing, these symbols, the applicable tolerance, and one or more datum symbols and modifiers are enclosed in a frame called a *feature control frame* (see Fig. 5-2). The standard geometric characteristic symbols and terms used in geometric tolerancing, as published in ANSI/ASME

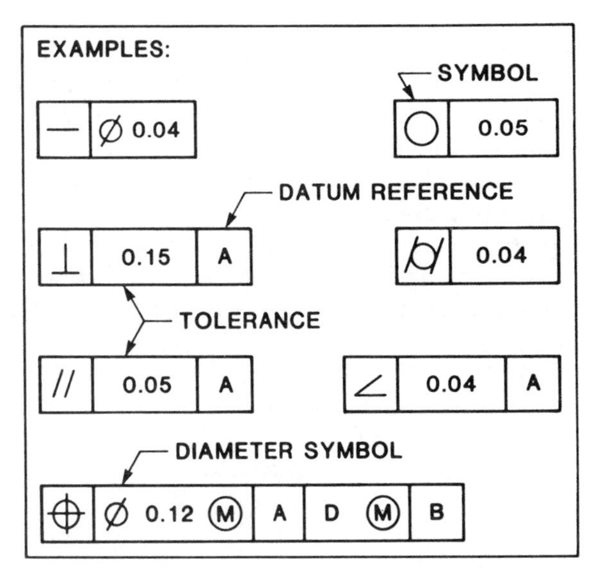

Figure 5-2 Feature control frames.

	CHARACTERISTIC	SYMBOL
FORM TOLERANCES	STRAIGHTNESS[1]	—
	FLATNESS[1]	▱
	CIRCULARITY (ROUNDNESS)	○
	CYLINDRICITY	⌭
PROFILE TOLERANCES	PROFILE OF A LINE[2]	⌒
	PROFILE OF A SURFACE[2]	⌓
ORIENTATION TOLERANCES	ANGULARITY	∠
	PERPENDICULARITY (SQUARENESS)	⊥
	PARALLELISM[3]	//
LOCATION TOLERANCES	POSITION	⌖
	CONCENTRICITY[3,7]	◎
	SYMMETRY[5]	⌖
RUNOUT TOLERANCES	CIRCULAR[4]	↗
	TOTAL[4,6]	⌰

[1]The symbol ∿ formerly denoted flatness. The symbol ⌒ or — formerly denoted flatness and straightness.

[2]Considered "related" features where datums are specified.

[3]The symbol ‖ and ⦿ formerly denoted parallelism and concentricity, respectively.

[4]The symbol ↗ without the qualifier "CIRCULAR" formerly denoted total runout.

[5]Where symmetry applies, it is preferred that the position symbol be used.

[6]"TOTAL" was formerly specified under the feature control frame.

[7]Consider the use of position or runout.

Figure 5-3 Geometric characteristic symbols

Y14.5M-1982 are given in Fig. 5-3. The meaning of the qualities conveyed by these symbols are:

Straightness specifies that any longitudinal element of a cylindrical part must lie between two parallel lines that are a distance apart equal to the specified tolerance.

Flatness specifies that all points of the actual surface must lie between two parallel planes that are a distance apart equal to the specified tolerance.

Circularity (or roundness) specifies the condition on a revolving surface (cylinder, cone, or sphere) wherein all points on the surface intersected by any plane are equidistant from the axis.

Cylindricity specifies the condition on a revolving surface wherein all elements form a perfect cylinder.

Profile of a line specifies the condition permitting a uniform amount of profile variation along a line element of a feature, either unilaterally or bilaterally.

Profile of a surface specifies the condition permitting a uniform amount of profile variation on a surface, either unilaterally or bilaterally.

Angularity specifies the condition of a surface or axis that is at some specified angle (other than 90 deg) from a datum plane or axis.

Perpendicularity (or squareness) specifies the condition of a surface, median plane, or axis that is at a 90 deg angle to a datum plane or axis.

Parallelism specifies the condition of a surface or axis that is equidistant at all points from a datum plane or axis.

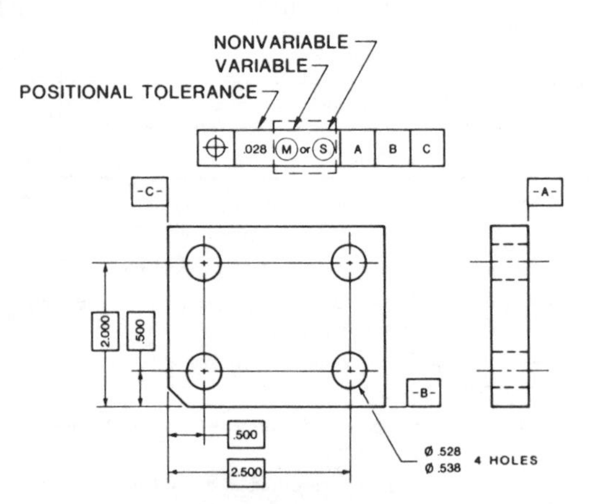

Figure 5-4 Positional tolerancing with feature control frame referenced to datum feature symbols.

Position specifies the true position, or theoretically-exact location of a point, line or plane (usually the center) of a feature in relation to a datum reference or other feature.

Concentricity specifies the condition wherein two or more features (cylinders, cones, spheres, hexagons) share a common axis.

Symmetry specifies the condition wherein either a part or feature has the same size and contour on both opposing sides of its median plane, or in which a feature shares a common plane with a datum plane.

Runout specifies the deviation from the desired form of the revolving surface of a part as detected during full rotation of the part on a datum axis when using a dial indicator or similar measuring device.

Applying Positional Tolerances

Several methods of indicating tolerances of location have evolved over the years. In the general tolerancing system, tolerances are applied to the locating dimensions of a drawing laid out according to conventional rectangular coordinate dimensioning. Conventional coordinate dimensioning is the method of rectangular dimensioning from surfaces or features established on the Cartesian coordinate system.

In this method, a part is drawn in relation to mutually-perpendicular reference planes (or rectangular or polar coordinates) called *base lines.* These are the reference lines or planes from which all dimensions are taken, using plus and minus tolerancing. The tolerances on part features—such as holes, surfaces, projections, slots, and other significant portions of a part—located by the conventional coordinate dimensions produce square, rectangular, or wedge-shaped tolerance zones. Complex hole patterns laid out with coordinate plus and minus tolerances not only can be interpreted different ways, but result in tighter tolerances, which often needlessly increase manufacturing costs.

Positional tolerancing applied to the Cartesian coordinate system, on the other hand, is more accurate than plus and minus tolerancing, because all references and tolerances related to basic dimensions and true position and are related to datums. This means that all coordinate dimensioning lines that establish the true position of holes, slots, bosses, and tabs in relation to plane and cylindrical surfaces, and that establish the center distances between such features, are "perfect" and can be interpreted *only one way.* Further, the positional tolerances specified in the feature control frame (see Fig. 5-4) for a hole describe a circular tolerance zone that substantially increases the target area for hole location, and which can reduce the cost of manufacture.

The feature control frame shown in Fig. 5-4 clearly indicates the datums from which the part features are dimensioned in their order of precedence reading left to right, the shape and diameter of the four tolerance zones, the relationship between the axes of the holes, and the diameter of the tolerance zones may be stated by *either* the MMC or RFS callout.

To summarize, in positional tolerancing, the untoleranced basic dimensions first locate the true geometrical center of a feature—axis, centerline, or center plane. Then, the positional tolerances are applied to true-position dimensions, allowing the center of the feature to lie anywhere within a tolerance zone of specified diameter or width. The shape of the tolerance zone reflects the geometry of the feature it locates. For example, the positional tolerance zone for an axial feature is a cylinder, within which the feature must lie (see Fig. 5-5).

Positional tolerancing is applied mainly to cylindrical features, but is also applied to other feature shapes, such as slots and tabs. It allows such features to be located without the use of compound tolerances, such as those described earlier in this Chapter under "Plus and Minus Tolerancing." Positional tolerancing used with the conventional coordinate dimensioning system to replace plus and minus tolerancing provides a more accurate and more economical way to produce parts that are truly interchangeable.

Inspecting Positionally-Toleranced Holes

Although all holes are located by two dimensions (usually at right angles to each other), a positional tolerance does not apply to these dimensions individually, but to the location of the center of the hole. You can visualize this tolerance as forming an imaginary circular zone of acceptability, having as its center the exact true position of the hole as located by basic dimensions, and the positional tolerance as a radius. If the center of the actual hole falls on or within this zone, the hole is acceptably located.

There are two classes of true hole positional tolerances: A and B. *Class A holes* have a positional tolerance of more than .002 in., and are frequently indicated by the note MMC. *Class B holes* have a positional tolerance of .002 or less and are frequently marked RFS.

You are allowed some latitude in the zone of acceptability when inspecting a Class A hole. Because the principles underlying the assembly of mating parts are based on MMC (which, in this case, is minimum hole size), you can subtract the minimum hole size specified on the drawing from the actual hole size as determined by inspection. The resulting difference may be added to the *diameter* of the zone of acceptability. If the actual hole center falls within the enlarged circle, the location of the hole is acceptable.

No such freedom is allowed with Class B holes, because the radius of the circle, or the positional tolerance, is nonvariable.

Having inspected the two location dimensions for the actual hole, you must determine the distance from true position to the actual hole center. This is, of course, the hypotenuse of a simple right triangle. The length of the hypotenuse will determine whether or not the hole is acceptably located. If it is less than the positional tolerance of a Class B hole, or less than the enlarged radius of the circle for that specific Class A hole, the location of the hole is satisfactory. Two common methods are used to find this distance: graphically, using the

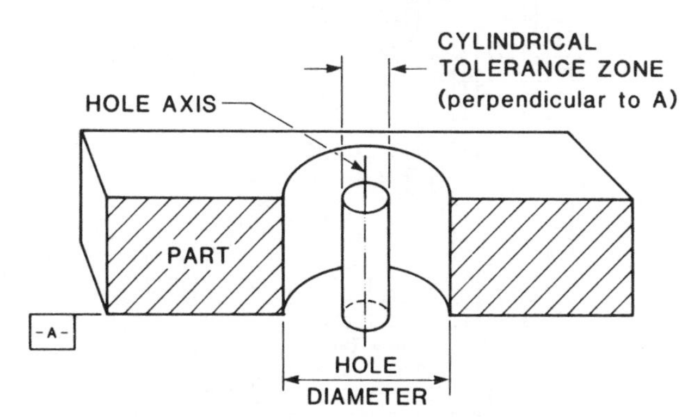

Figure 5-5 The positional tolerance zone for an axial feature is a cylinder.

target-type graph illustrated in Fig. 5-6, or solving a simple geometric or trigonometric problem. The first method involves considerable paperwork. The second method may be preferable, especially in a training situation. Several examples follow and more may be easily devised by the instructor.

Graphic solution: A .250 ± .001 dia hole, located within .002 of true position. Dimension location is 1.500 left of V℄, .750 up from H℄. Inspection shows actual dimensions to be 1.5014 and .7516. The graph indicates by the intersection of the two dimensions that the actual center falls outside the .002 circle and, therefore, that the hole is not acceptable.

Mathematic Example No. 1: A .375 ± .005 dia hole, located within .005 of true position. True position (TP) location is .395 left of V℄ and .415 up from H℄. Inspection shows that the actual hole size is .376, and that actual location dimensions are .418 and .397.

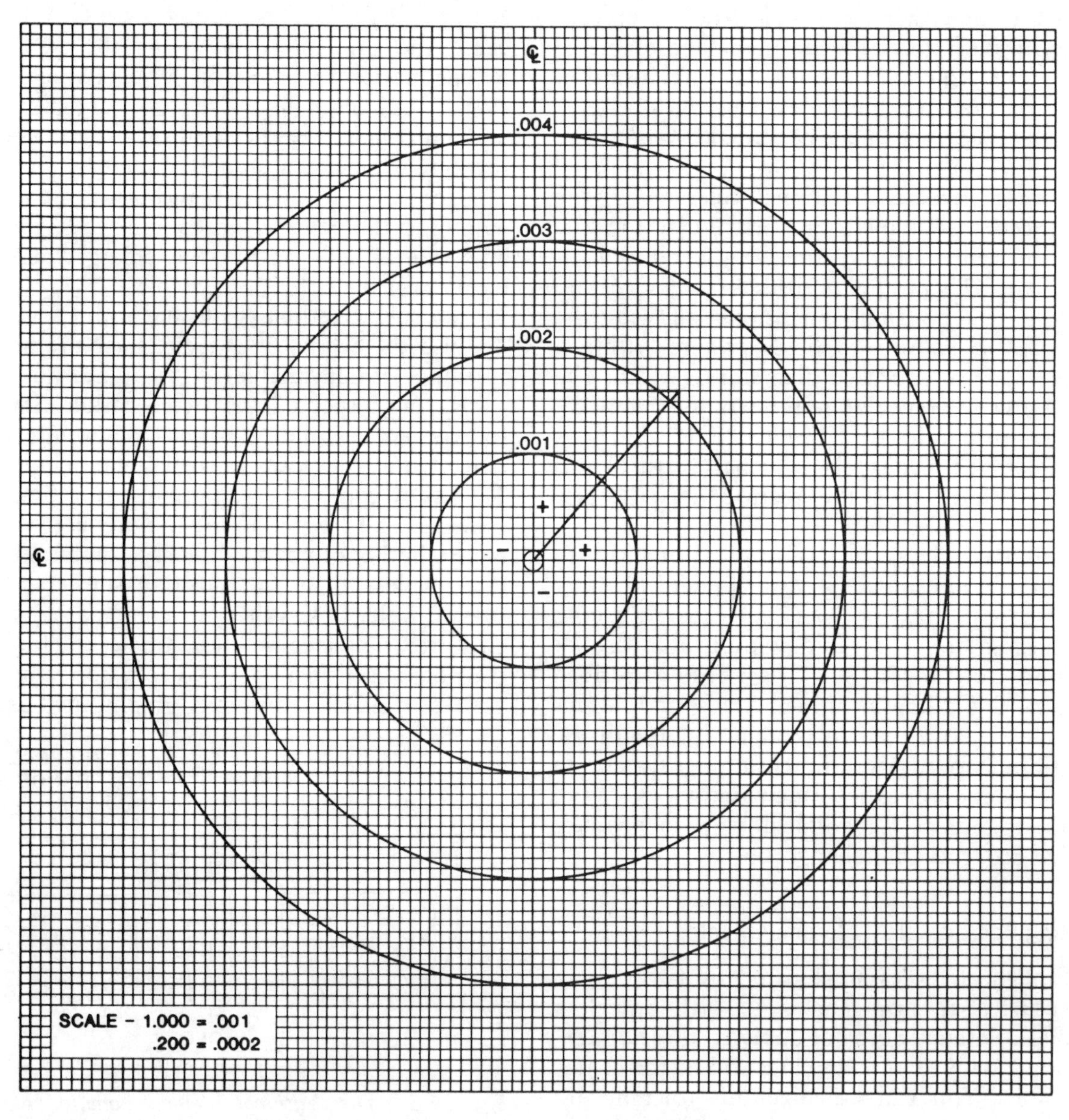

Figure 5-6 Target for inspecting positionally-toleranced Class A and Class B holes.

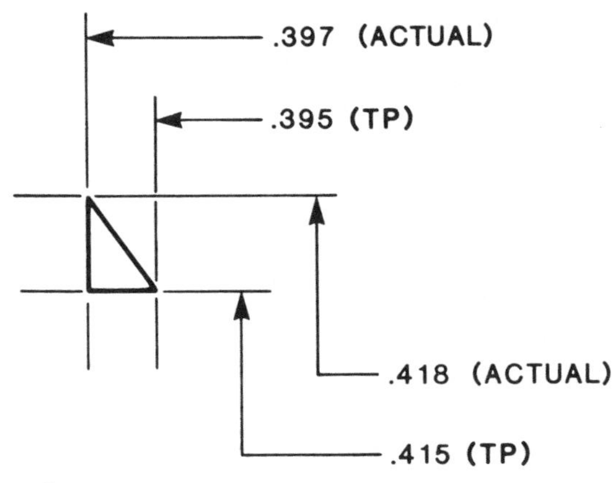

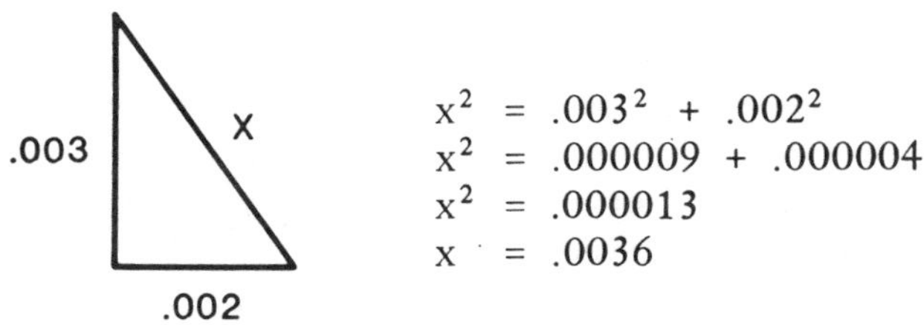

$$x^2 = .003^2 + .002^2$$
$$x^2 = .000009 + .000004$$
$$x^2 = .000013$$
$$x = .0036$$

Determining circle dia:
(Class A hole)

$$D = 2\,(\text{TP tol}) + (\text{Actual dia} - \text{Min dia})$$
$$D = 2\,(.005) + (.376 - .375)$$
$$D = .010 + .001$$
$$D = .011$$
$$R = \frac{.011}{2} = .0055$$

Because .0036 is less than .0055, the hole is acceptably located.

Mathematic Example No. 2: A .250 dia hole, $^{+.000}_{-.005}$ located within .005 of true position. True position dimension location is .375 left of V℄ and 1.425 up from the H℄. Inspection shows that the actual hole size is .247 and that actual location dimensions are 1.4295 and .3708.

Determining circle dia:
(Class A hole)

$$D = 2\,(.005) + (.247 - .245)$$
$$D = .010 + .002$$
$$D = .012$$
$$R = \frac{.012}{2} = .006$$

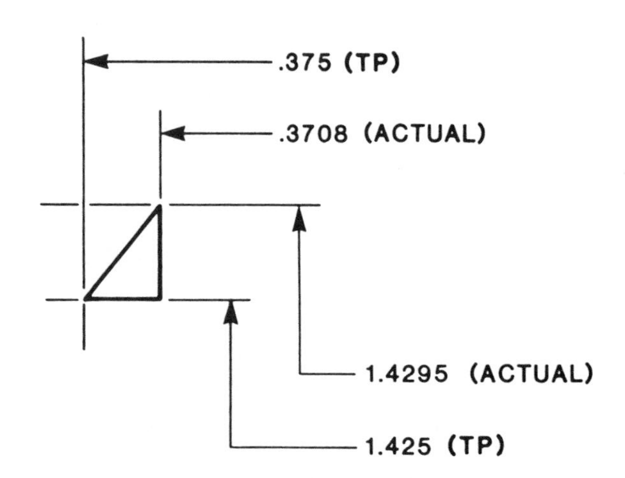

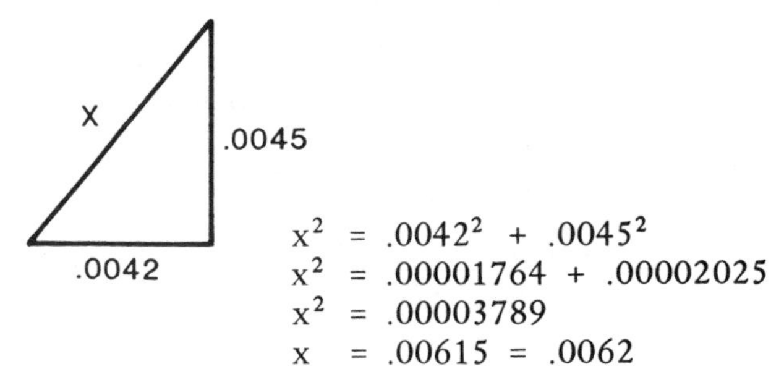

$$x^2 = .0042^2 + .0045^2$$
$$x^2 = .00001764 + .00002025$$
$$x^2 = .00003789$$
$$x = .00615 = .0062$$

Because R is less than x, the hole is not acceptable. However, one of the advantages of a Class A hole is its variability, which allows a hole not located exactly to be "saved" within a narrow range. Using Mathematical Example No. 2, the radius of the zone of acceptability is .006 and the hypotenuse is .0062, putting the hole outside limits. But, if we enlarge the hole .003, taking it to its maximum size of .250, the radius becomes .015 [2 (.005) + (.250 - .245)]. This gives a radius of .0075, which is greater than .0062. The hole is now acceptable. Sometimes even greater deviations can be salvaged, if the hole size permits enlarging and if the hole center can be simultaneously shifted, although this calls for careful machining.

SURFACE TEXTURE

The demand has grown in recent years for a more exact method of specifying surface roughness than simply marking the drawing "finish," "grind," or "lap." Not only the service life, but also the proper functioning of a part may depend on obtaining the smoothness quality required for contact surfaces. Figure 5-7 illustrates the characteristics associated with surface texture.

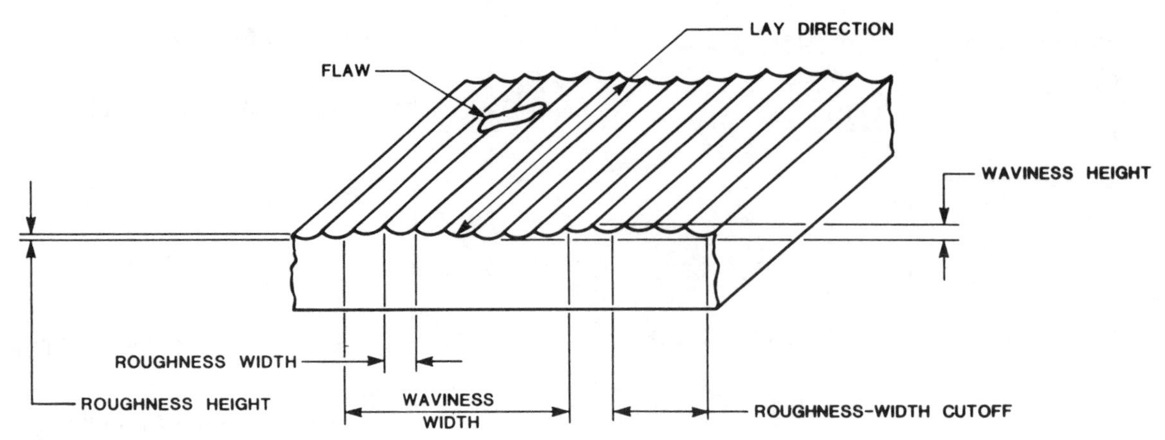

Figure 5-7 Surface texture characteristics.

Surface texture is defined as the repetitive or random deviations from the nominal surface that form the pattern of the surface. These include: roughness, waviness, lay, and flaws.

Roughness describes the relatively finely-spaced surface irregularities produced by the cutting action of tool edges and abrasive grains on machined surfaces.

Roughness height is a number, given in microinches (.000001 in.) representing the arithmetic mean of approximately five profilometer readings. (A *profilometer* is a device that measures the surface characteristics of a material.) Roughness height is the most important surface characteristic to a machinist.

Roughness width is the distance between adjacent peaks or ridges that constitute the predominant pattern of roughness. Roughness width is measured in inches or millimeters.

Waviness describes the surface undulations or waves that may result from machine or workpiece deflections, vibrations, warping, strains, etc.

Waviness height is its peak-to-valley distance measured in inches or millimeters.

Waviness width is the space between adjacent wave valleys or wave peaks measured in inches or millimeters.

Flaws are irregularities, such as cracks, checks, blowholes, scratches, etc., that occur on a surface.

Lay is the predominant direction of the tool

Table V-1: Roughness Height Values.

(Microinches)	*Type of Surface*	*Purpose*
1000	Extremely rough	Clearance surfaces where appearance is not important
500	Rough	Used where stress requirements and close tolerances are not required
250	Medium	Most popular where stress and tolerance requirements are essential
125	Average smooth	Suitable for mating surfaces and parts held by bolts and rivets with no motion between them
63	Better than average finish	For close fits or stress parts, except rotating shafts, axles, and parts subject to extreme vibration
32	Fine finish	Used where stress concentration is high, and for applications such as bearings
16	Very fine finish	Used where smoothness is of primary importance, i.e., high-speed shaft bearings, heavily loaded bearings, and extreme tension members
8	Extremely fine finish by grinding, honing, lapping or buffing	Used for cylindrical surfaces

marks that produce a surface pattern.

Surface Quality Symbols

The machining industry uses two types of symbols to indicate surface quality or texture: controlled and non-controlled.

Controlled symbols are illustrated in Fig. 5-8. The construction of the symbol is based on a check mark crossed at the top with a short horizontal line. Except for lay and roughness height, surface characteristics are designated by measurements. The degrees of finish assigned to the various roughness heights are given in Table V-1.

Non-controlled symbols tell little more than that a given surface requires finishing. Occasionally such symbols (see Fig. 5-9) reference notes that provide more information. In addition to the symbols, the note FOA is used, which stands for "Finish all over." This means that all surfaces must receive a machine finish.

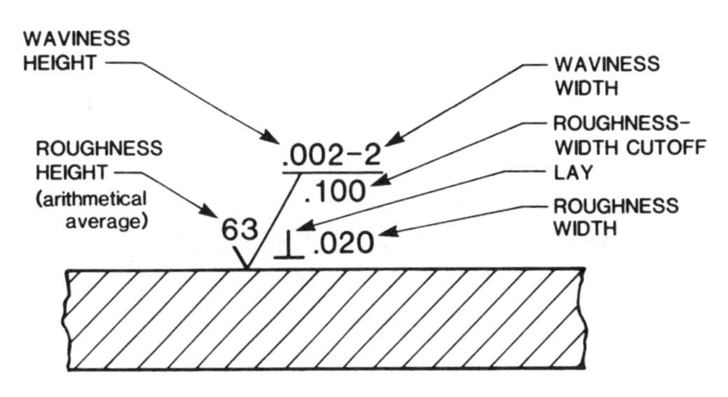

Figure 5-8 Controlled symbols for surface quality.

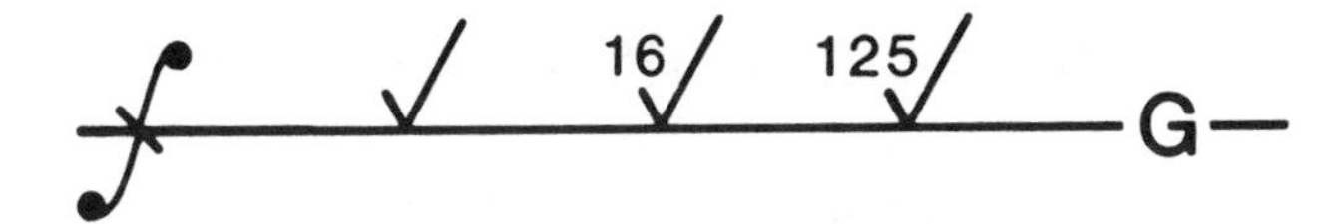

Figure 5-9 Non-controlled symbols for surface quality.

SUMMARY

Geometric dimensioning and tolerancing applies realistically to the demands of modern manufacturing. Drawings utilizing these techniques are easier and faster to read than similar drawings that rely on complete notes. Individual features are toleranced for their acceptable variations in size, position, and form.

Blueprints depict surface quality in two ways, by controlled and non-controlled symbols. Controlled symbols cover every aspect of surface finish, including roughness, waviness, lay, etc. Machinists are most often concerned with roughness. Roughness is measured in microinches using a device called a profilometer.

TRAINING PRACTICE

Blueprint No. 5

1. Give the name of the part. ______________________________

2. What type of material is called for? ______________________________

3. Describe the material and its composition (Hint: use the *Machinery's Handbook* for description).

____

____

4. How many datums are shown on this print? ______________________________

5. What requirement is made of Ⓑ with reference to [A]? ______________________________

____

6. What requirement is made of Ⓒ with reference to [A]? ______________________________

____

7. What kind of requirement is placed on the .500 dia. hole? ______________________________

____

What does the symbol [Ø] mean? ______________________________

____

8. Describe the characteristics of a 16 microfinish. ______________________________

____

____

9. What does the "16" in a 16 microfinish mean? ______________________________

____

10. What is the maximum permissible size of the 3.000 dia? ______________________________

____

11. What does the [⌭] symbol mean? ______________________________

____

12. What is the minimum permissible length of the part? ______________________________

____

13. What is the tolerance on two-place decimals: ____________________

____

14. What machining operation will yield a 32 microfinish? ____________________

____

15. How can a 16 microfinish be obtained? ____________________

____

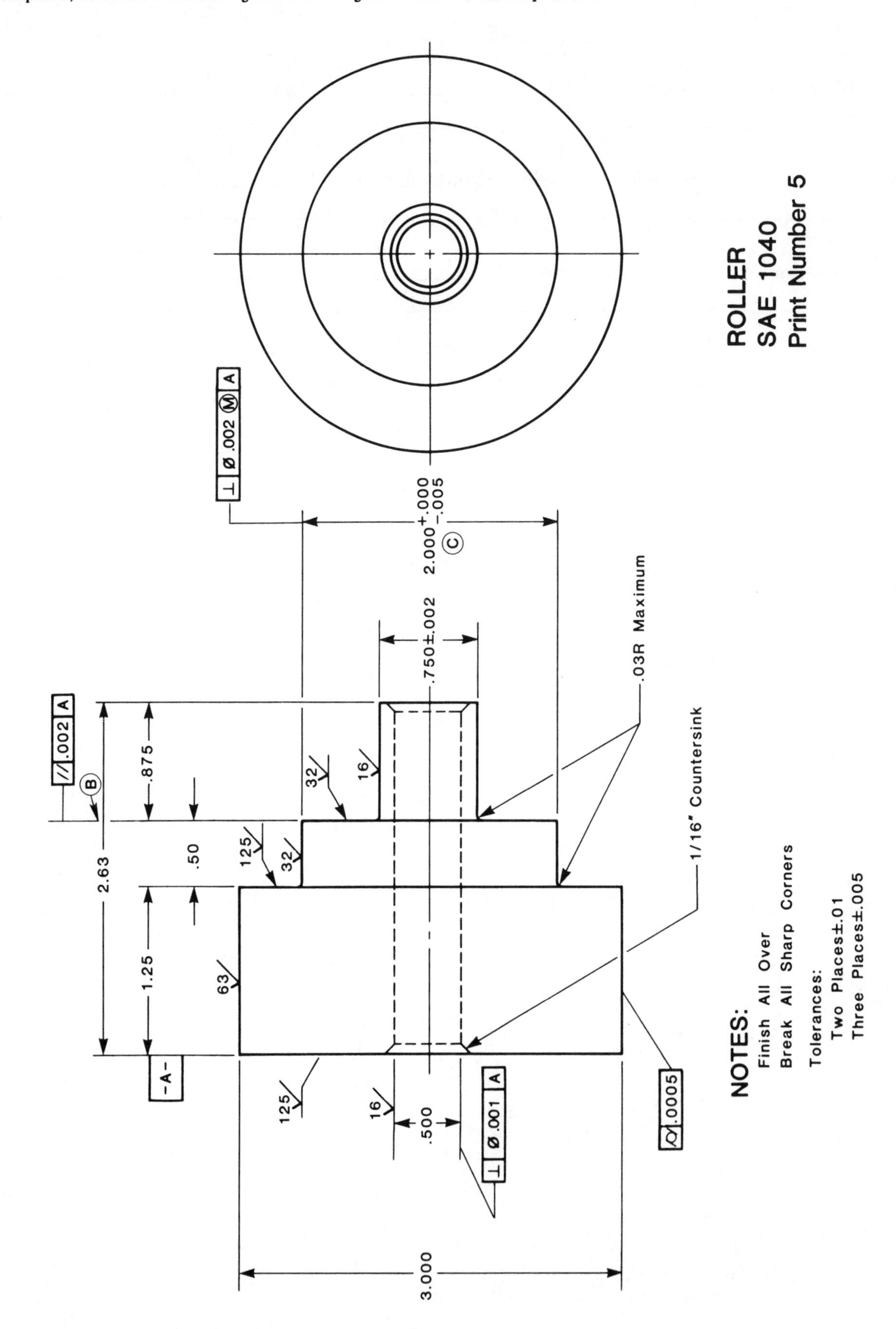
ROLLER
SAE 1040
Print Number 5
NOTES:
Finish All Over
Break All Sharp Corners
Tolerances:
Two Places±.01
Three Places±.005
.03R Maximum
1/16" Countersink
2.000 +.000 −.005
C
.750±.002
.875
2.63
.50
1.25
3.000
.500
-A-
B
⊥ Ø .002 M A
// .002 A
⊥ Ø .001 A
.0005
125
32
16
63

Chapter VI

METRICS AND METRIC DRAWINGS

The metric system is an international language of measurement that is gradually being adopted by the manufacturing industries in the United States. Its formal name is the *International System of Units* (or *Le Système International d'Unités* in the French language), which is officially abbreviated "SI" in all languages. SI is a modern version of the metric system developed in France during the French Revolution, adopted in 1960 by the General Conference on Weights and Measures. It provides a logical and interconnected framework for all measurements in science, industry, and commerce that allows expanded exchange of industrial goods among the nations of the world. Therefore, the use of metric units in U.S. manufacturing is increasing, and gradually replacing our customary inch (or English) units of measurement.

The transition from inch unit dimensions to metric unit dimensions on engineering drawings is occurring in both directions between metric and non-metric countries. The frequent exchange of product designs makes it necessary to convert some basic inch designs into equivalent millimeter designs, and *vice versa.* Thus, it is very important for you to understand the metric system.

SI BASE UNITS

The SI system has three classes of units: base units, supplementary units, and derived units. The seven base units are listed in Table VI-1. You do not need to concern yourself with supplementary and derived units at this time.

Table VI-1: SI Base Units.

Physical Quantity Measured	*Name of Base SI Unit*	*SI Symbol*
Length	meter	m
Mass	kilogram	kg
Time	second	s
Electric current	ampere	A
Thermodynamic temperature	kelvin	K
Amount of substance	mole	mol
Luminous intensity	candela	cd

Multiple and Sub-Multiple Prefixes

Prefixes are added to the base unit names and symbols to form the names and symbols of the multiples and sub-multiples of the SI units, as given in Table VI-2. The same prefixes apply to any SI base unit. The multiples and sub-multiples reflect the powers of 10. Recall that, in mathematics, we speak of *base* and *power.* For example, in 3^5, *3* is the base and *5* is the exponent. Thus, the expression reads: "3 to the 5th power." Because metric units are based upon the decimal system of counting or multiples of 10, the multiples or sub-multiples of any unit are related to the unit by powers of 10.

As shown in Table VI-2, sets composed of ones, tens, and hundreds are grouped into "periods" of three digits, separated by a space or a decimal point (instead of a comma as in the inch system). Ones can occupy both the first and last position in a set of three digits. Tens are always in the second position, and hundreds are always in

Table VI-2: SI Prefixes for Multiples and Sub-Multiples (Powers of Ten).

Numerical Magnitude	*Powers of 10*	*Prefix*	*Symbol*
1 000 000 000 000	= 10^{12}	tera	T
1 000 000 000	= 10^{9}	giga	G
1 000 000	= 10^{6}	mega	M
1 000	= 10^{3}	kilo	k
100	= 10^{2}	hecto	h
10	= 10^{1}	deka	da
0.1	= 10^{-1}	deci	d
0.01	= 10^{-2}	centi	c
0.001	= 10^{-3}	milli	m
0.000 001	= 10^{-6}	micro	μ
0.000 000 001	= 10^{-9}	nano	n
0.000 000 000 001	= 10^{-12}	pico	p

the third position. Therefore, whether you read from left to right or right to left from a space or decimal point, the digit that is one place over is a multiple or sub-multiple of 1, the digit two places over is a multiple or sub-multiple of 10, and the digit three places over is a multiple or sub-multiple of 100, etc.

The Meter—SI Basic Unit of Length

Because we are concerned only with dimensions on drawings in this course, this Chapter will deal only with metric units of length. As shown in Table VI-1, the basic SI unit for measuring length is the *meter,* whose symbol is a lower-case m. Applying the prefixes given in Table VI-2 to the meter, the multiple and sub-multiples of the meter are shown in Table VI-3. The millimeter (mm) is used almost exclusively for part dimensions on drawings. Using the millimeter as the standard dimension eliminates the need to label each dimension. Just as inch dimensions are described on drawings as so many "thousandths" and "tenths" without actually adding the word "inches," the equivalent range of millimeter dimensions are

Table VI-3: Multiples and Sub-Multiples of the Meter.

10^{-3} meter	=	a *milli*meter	(mm)	.001
10^{-2} meter	=	a *centi*meter	(cm)	.01
10^{-1} meter	=	a *deci*meter	(dm)	.1
10^{0} meter	=	a meter	(m)	1
10^{1} meters	=	a *deka*meter	(da)	10
10^{2} meters	=	a *hecto*meter	(ha)	100
10^{3} meters	=	a *kilo*meter	(km)	1000

described as "tenths" or "hundredths" without adding the word "millimeters."

Converting English Units to Metric Units

The following conversions will help you mentally compare the English inch units with metric units:

1 in = 25.4mm
1 ft = 30.48cm
1 yd = 0.9144m

Because, as you can see, the millimeter is so much smaller a unit of measure than the inch, "tenths" and "hundredths" of millimeters are used frequently. "Thousandths" of a millimeter are used less frequently, and then only on those parts where "hundredths" or "thousandths" of inches are common. The "half millimeter" (0.5mm) is also in common use, because it is equal to about 0.020 in., and close to 1/65 in.

The Metric Rule

The half millimeter (0.5mm) is the smallest graduation found on steel rules. However, most ordinary metric rules are divided into only the millimeter (1/1000th of a meter). Numerals are printed on the rule at every 10mm mark, as shown in Fig. 6-1.

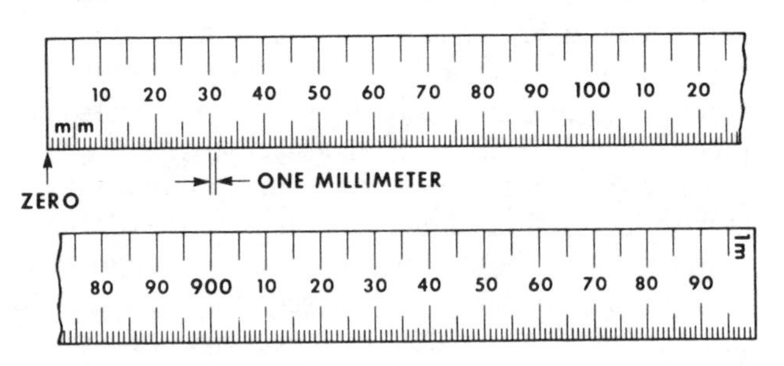

Figure 6-1 There are 1 000mm in one meter, or 1 000mm = 1m.

Important Metric Guidelines

As in any system, there are standardized guidelines to follow, and metrics is no exception. The most important ones for you to remember when using metric dimensional units are:

(1) Identify all symbols for metric units properly.

For example, always use the lower case "m" for meter, "cm" for centimeter, and "mm" for millimeter. Never use capital letters, because they may change the meaning.

(2) Do not use commas or periods between sets of three whole digits. Use a space instead. For example, 1 000mm. This rule applies anytime you have more than three whole digits, as shown in Table VI-2.

(3) Use the decimal point (not a space) to indicate fractional digits, and add a zero (0) before a fractional digit that is not preceded by a whole digit (however, this practice is not always followed on blueprints).

(4) Do not use a space between a digit and the symbol for a metric unit. For example, 4mm, 8m, etc. Do not use a space between the symbol for a prefix and the symbol for an SI basic unit. For example, 4mm, 5cm, etc.

(5) Always use the identifying symbol with each digit. For example, 4mm, 8m, etc., except on a metric drawing where the units of measure applied to dimensions are indicated by standardized conventions (see below).

(6) Apply only one prefix at a time to a basic unit. For example, 0.1 centimeter is not called a deci-centimeter, but a millimeter.

TYPES OF METRIC DRAWINGS

You will encounter three types of metric drawings in a manufacturing/fabricating environment: dual dimensioned drawings, conversion chart metric drawings, and all-metric drawings, each of which has different dimensioning characteristics that you must know.

Dual Dimensioned Drawings

Dual dimensioning (see Fig. 6-2) means using

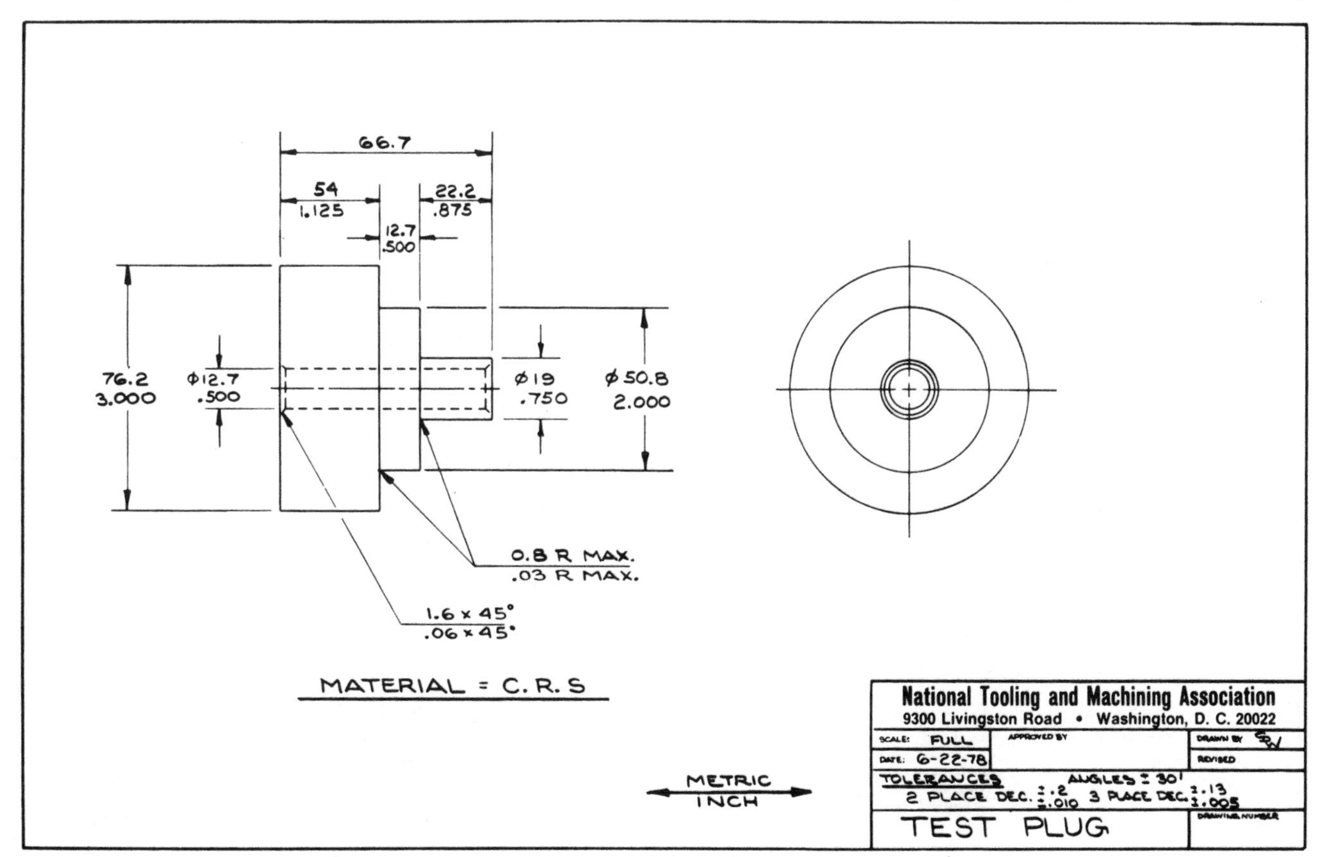

Figure 6-2 A dual dimensioned metric drawing.

two different units of measurement on the same drawing to show geometric form and shape. Dual dimensioned metric drawings give dimensions in both the inch and the millimeter. These units may be located within a dimension line in one of four different ways:

(1) Stacking the inch dimension directly above the metric dimension, separated by a horizontal line, as:

$$\frac{1.246 \pm .004 \quad \text{in.}}{31.65 \pm .1 \quad \text{mm}}$$

(2) Similar to (1), except that the metric dimension is stacked above the inch dimension, as:

$$\frac{31.65 \pm .1 \quad \text{mm}}{1.246 \pm .004 \quad \text{in.}}$$

(3) Enclosing the metric dimension with square brackets, while leaving the inch dimension open, as:

[31 65 ± .1] 1.246 ± .004
mm in.

(4) The opposite of (3). The inch dimension is enclosed in brackets, and the metric dimension is left open, as:

[1.246 ± .004] 31.65 ± .1
in. mm

A special note, appearing adjacent or close to the title block, will state the location of the inch unit with respect to the millimeter unit, such as:

$$\frac{\text{millimeter}}{\text{inch}} \qquad \frac{\text{inch}}{\text{millimeter}}$$

[millimeter] inch [inch] millimeter

Of the four methods of dual dimensioning, numbers (2) and (3) above are the most common on drawings, because they are industry-accepted formats. However, one of the other two formats

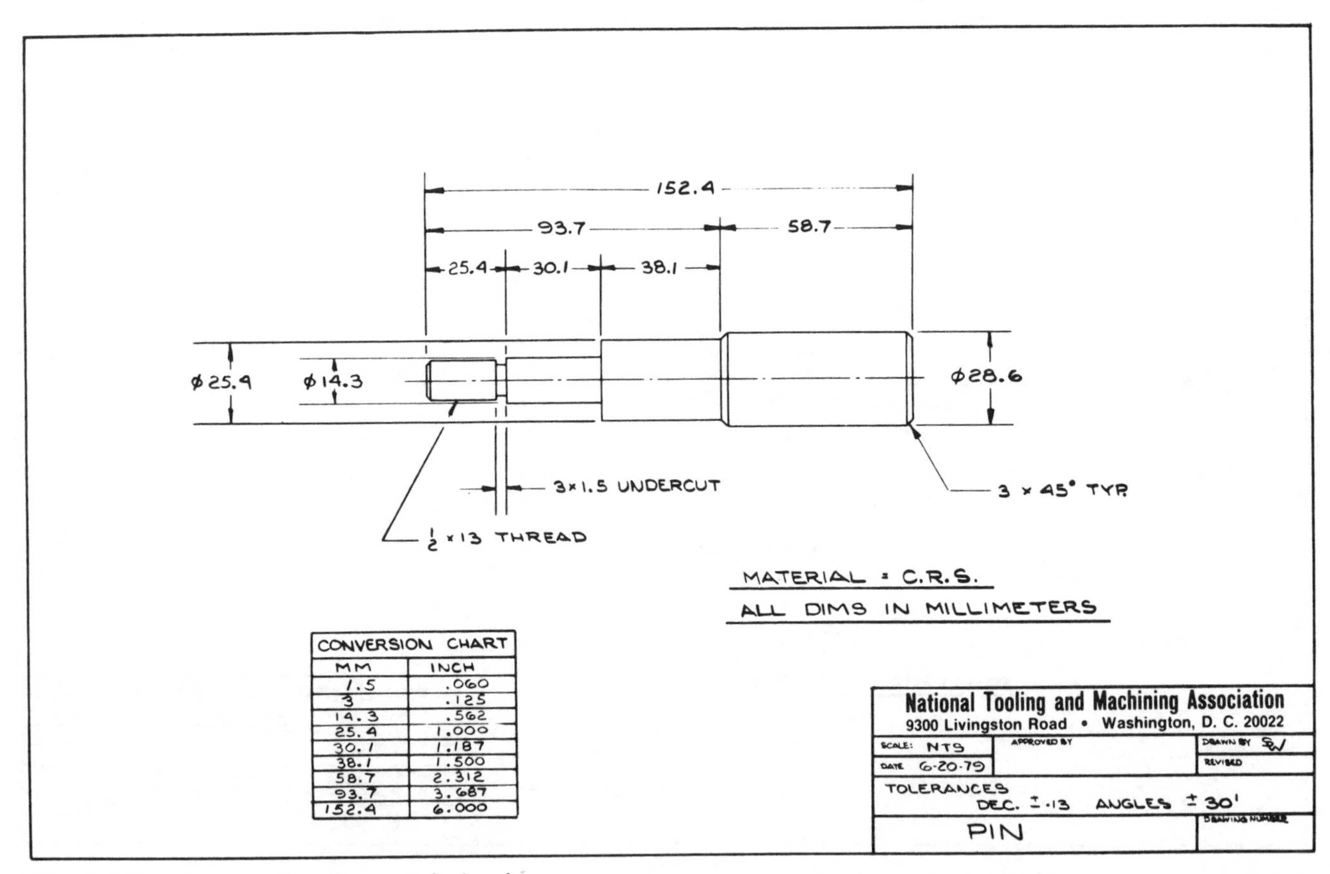

CONVERSION CHART	
MM	INCH
1.5	.060
3	.125
14.3	.562
25.4	1.000
30.1	1.187
38.1	1.500
58.7	2.312
93.7	3.687
152.4	6.000

Figure 6-3 A conversion chart metric drawing.

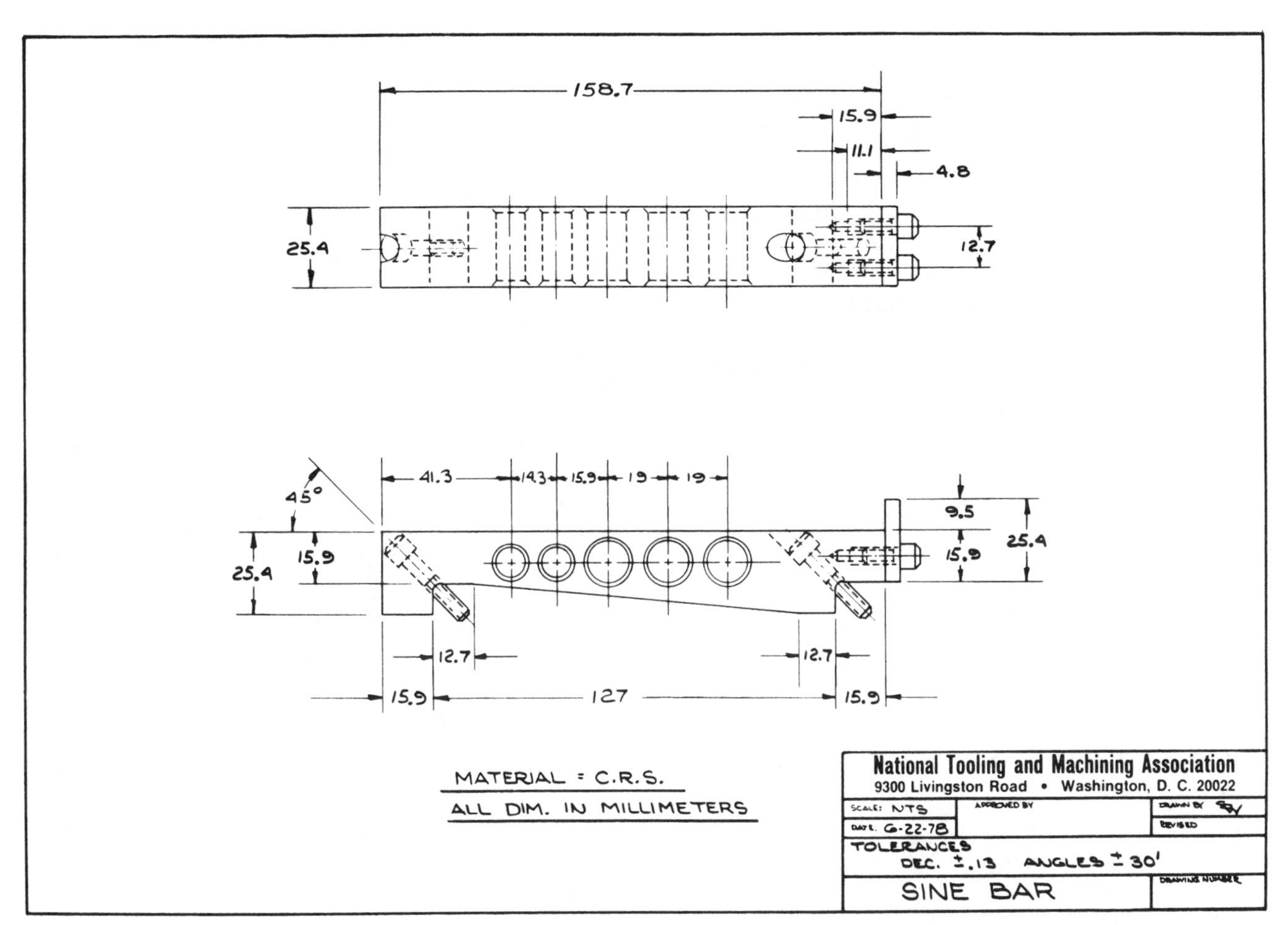

Figure 6-4 An all-metric drawing.

may be used, depending upon a particular company's engineering and design practice. When working with dual dimensioned drawings, be very careful not to confuse the two different units of measurement with one another, because there is the definite possibility of making the part either larger or smaller than required if you do. Misinterpreting 25.4mm for 25.4 in. would cause a part to be approximately 25 times larger than required.

Conversion Chart Metric Drawings

Conversion chart metric drawings differ from dual dimensioned drawings in that only one unit of measurement–either metric or inch–is shown within the dimension line, and the counterpart dimensional unit is given in a conversion chart located on the drawing. When an dimensions are in millimeters, the conversion chart will give the direct conversion of the metric units to inch units as shown in Fig. 6-3.

Soft Conversion. Both dual dimensioned and conversion chart metric drawings are called *soft converted* metric drawings, because all specifications conform to U.S. standards. No special tooling, such as end mills, drills, reamers, taps, etc., are required unless noted otherwise. Conventional, standard decimal/inch tooling can be used to manufacture the part.

All-Metric Drawings

In an all-metric drawing (see Fig. 6-4), every dimensional feature is in strict conformance with standard *metric* practice, and there is no direct conversion to the inch unit on the drawing. In other words, it is "all metric." This type of drawing is in strict conformance to SI standards, including all specifications, such as surface texture or finish, screw threads, heat treatment, etc.

Hard Conversion. All-metric drawings are called *hard converted* drawings, because the design and engineering is based on SI metric standards. This means that conventional U.S. tooling, such as end mills, reamers, drills, etc., along with gaging and measuring devices, cannot be used to manufacture the part. In addition, machine tools having metric capabilities are required to machine dimensional features, and metric measuring devices—such as micrometers, depth gages, and machinist's scales—are required in the manufacturing process.

FIRST-ANGLE PROJECTION

Engineering drawings developed in Europe use first-angle projection (see Fig. 6-5), as you

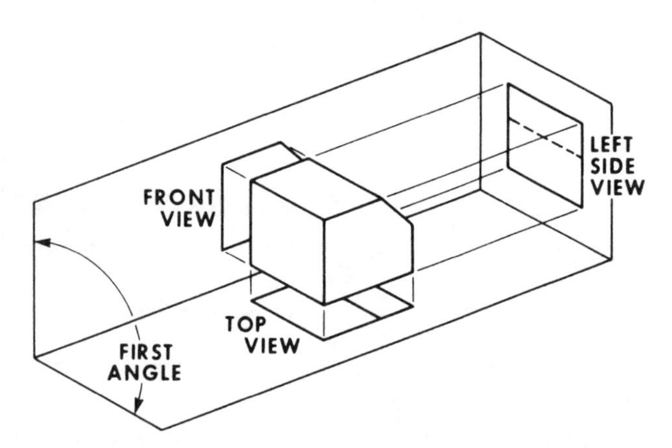

Figure 6-5 First angle projection of a cube (compare with Fig. 2-5).

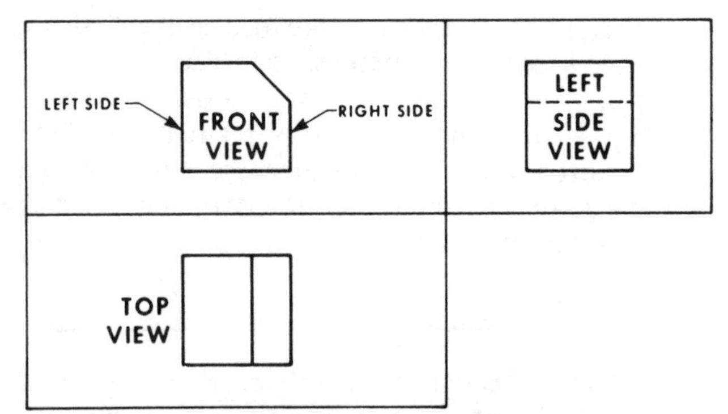

Figure 6-6 First angle projection brought into a single plane.

learned in Chapter II. First-angle projection is considered illogical in the United States because, when the three views are rotated into the same

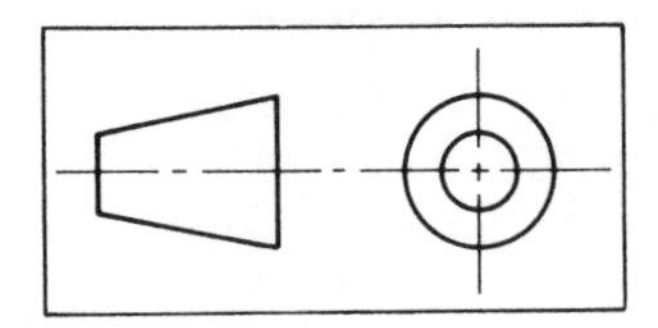

Figure 6-7 Symbol for first angle projection.

plane, the top view is *below* the front view, and the left side of the object is to the *right* of the front view, as shown in Fig. 6-6. Drawings using first-angle projection indicate this by the graphic symbol shown in Fig. 6-7.

METRIC DRAWING IDENTIFICATION

In all-metric drawings, close to the title block/specification block, you will see a special note stating that all dimensions and surface finishes given on the drawing are in millimeters, as shown in Figs. 6-3 and 6-4. Sometimes the word "Metric" is printed or stamped on the drawing.

METRIC CONVERSION FACTORS

When converting from inches to millimeters and millimeters to inches you must follow formal conversion guidelines. The reasons for this are to maintain the accuracy of the original unit of measurement, and to insure the interchangeability of the part.

Converting Dimensions without Tolerance

To convert inch dimensions without a tolerance to millimeters, carry the millimeters to one (1) additional significant digit *if the first digit of the converted millimeter value is the same as, or less than, the first digit of the inch value.*

Example: Convert 5.4 in. (two significant digits) to mm.
1" = 25.4mm

Therefore:
5.4 in.
x 25.4mm
= 137.16mm

Round 137.16mm to 137mm (three significant digits).

When the first digit of the millimeter value is greater than the first digit of the inch value, carry the millimeter value to the same number of significant digits as the inch value.

Example: Convert 2.3 in. (two significant digits) to mm
1″ = 25.4mm

Therefore:

$$\begin{array}{r} 2.3 \text{ in.} \\ \times\ 25.4\text{mm} \\ \hline = 58.42\text{mm} \end{array}$$

Round 58.42mm to 58, because the first digit of the millimeter value (5) is greater than the first digit of the inch value (2). Thus, you carry the millimeters to the same number of significant digits as the inch dimension (two).

Converting Tolerances

There are two methods for rounding off converted tolerances.

Method A:

(1) Calculate the maximum and minimum limits in inches.

(2) Convert the values in Step 1 exactly to millimeters by multiplying the inch values by 25.4mm.

(3) Round the results to the nearest value as given in Table VI-4, according to the original tolerance in inches.

Example: Convert 1.950 ± 0.016 in. to mm

(1) Determine limits
Upper: 1.950 + 0.016 = 1.966 in.
Lower: 1.950 − 0.016 = 1.934 in.

(2) Convert the upper and lower limits to mm
Upper: 1.966 x 25.4mm = 49.9364mm
Lower: 1.934 x 25.4mm = 49.1236mm

(3) Round 49.9364 to 49.94mm
Round 49.1236 to 49.12mm

In Step (1) under the example, we determined the upper and lower limits of 1.934 and 1.966 in. They equalled a total limit range of 0.032 in. (1.966 − 1.934 = 0.032). Referring to Table VI-4 under the heading, "Original Tolerance Range (in.)," the 0.032 value is at least 0.01 and less than 0.1. Therefore, we can round the metric limits to two places, or 0.01mm. Thus, the upper limit of 49.9364mm can be rounded to 49.94mm, and the lower limit of 49.1236mm can be rounded to 49.12mm.

Using Method A, metric conversion remains statistically accurate, even though the converted limits may vary up to 2% of the tolerance. Therefore, when using this method, the converted values must be used as the basis for inspection.

Method B is used to convert tolerances that are extremely critical to the manufacturing of parts and components. This method is used where mating parts and original inspection equipment are concerned.

Method B:

(1) Determine the upper and lower limits in inches.

(2) Determine the total tolerance limit by subtracting the lower limit from the upper limit.

(3) Reference Table VI-4 to determine which decimal place should be used for rounding.

(4) Convert the upper and lower inch limits to mm (multiply by 25.4mm).

(5) Round the upper limit to the next lower value, and round the lower limit to the next higher value.

Example: Convert 1.950 ± .016 in. to mm

(1) Determine limits
Upper: 1.950 + .016 = 1.966 in.
Lower: 1.950 − .016 = 1.934 in.

(2) Find the total tolerance limit:

$$\begin{array}{r} 1.966 \\ -\ 1.934 \\ \hline = 0.032 \text{ in.} \end{array}$$

(3) Using Table VI-4, we find that 0.032 is at least 0.01 and less than 0.1 in the "Original Tolerance Range (in.)" column, and the "Round to (mm)" column indicates round-

Table VI-4: Rounding Tolerances (in. to mm).

Original Tolerance Range (in.)		*Round to (mm)*	
at least	*less than*		
0.000 01	0.0001	0.000 01	5 decimal places
0.0001	0.001	0.0001	4 decimal places
0.001	0.01	0.001	3 decimal places
0.01	0.1	0.01	2 decimal places
0.1	1.0	0.1	1 decimal place

ing to two decimal places, or 0.01mm.

(4) Convert the upper and lower inch limits to mm
Upper: 1.966 x 25.4mm = 49.9364mm
Lower: 1.934 x 25.4mm = 49.1236mm

(5) Rounding the upper limit to the next lower value of hundredths decimal place yields 49.93, and rounding the lower limit one digit higher than the hundredths decimal place yields 49.13. Therefore, the new upper metric limit is 49.93, and the lower limit is 49.13.

In summary, careful conversion and rounding of tolerances when converting to metric units is extremely important. If accuracy is lost in the metric conversion, parts will not function properly and interchangeability will be lost.

SURFACE TEXTURE

In the inch system, surface texture quality is measured in microinches, and is indicated by the symbol $\sqrt{}$. This same symbol is used in the metric system. The major difference when specifying surface texture in the metric system is that the micrometer (μm) is the unit of measurement instead of the microinch. The metric value given with the symbol states the type of finish required on a machined part. Table VI-5 lists the most common surface finish classes in both inches and their respective metric counterparts.

Table VI-5: Surface Finish.

Microinch	*Micrometer*
.5	0.012
1	0.025
2	0.05
4	0.1
8	0.2
16	0.4
32	0.8
63	1.6
125	3.2
250	6.3
500	12.5
1000	25.0
2000	50.0

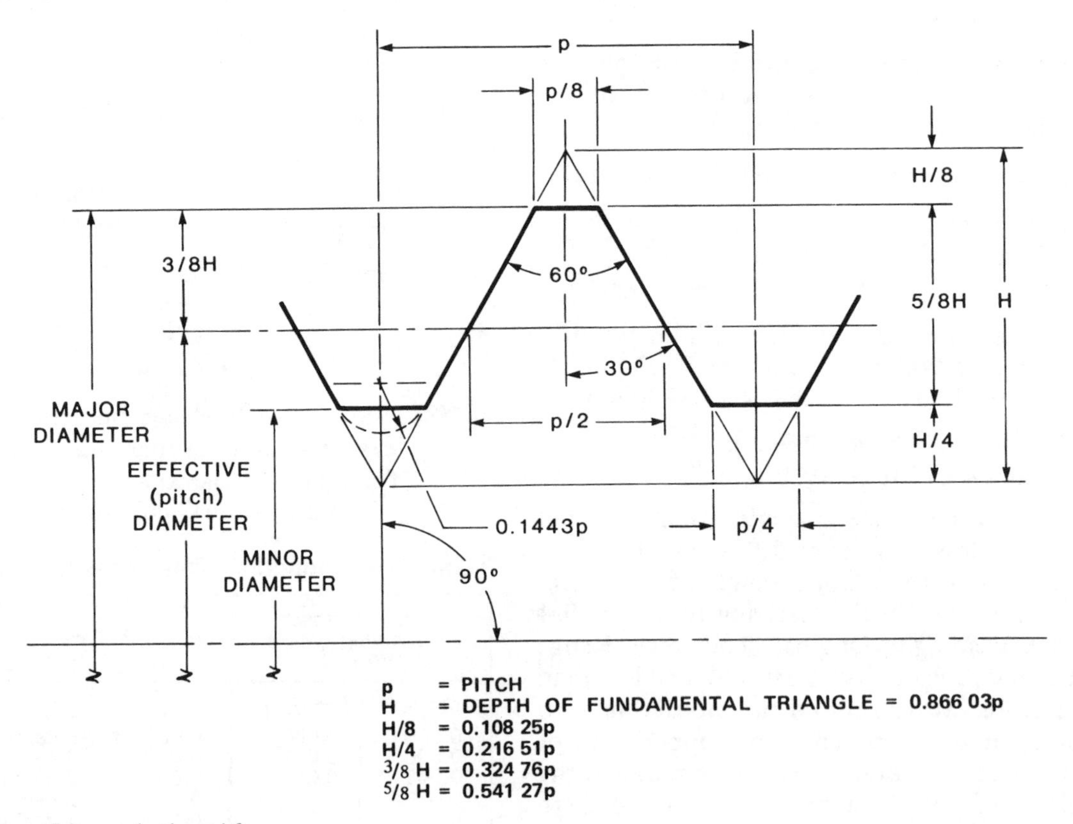

Figure 6-8 ISO metric thread form.

METRIC SCREW THREAD NOTATION

The metric screw thread form (see Fig. 6-8) is similar to the American and Unified forms. Each has a 60 deg included angle and similar flat sizes at their crests. However, the flat at the root is larger on metric threads.

The metric system of screw thread notation is similar to the Unified system described in Chapter IV, with some variations. All units of measure are given in millimeters, and the metric system has only three thread series: Coarse, Fine, and Extra-Fine. Coarse threads have emerged as being satisfactory for general application. There are no number-size metric threads like the 8, 12, and 16 Series in the Unified system.

Whereas Unified thread notation states major diameter as either a number or a dimension followed by the TPI, ISO metric threads are designated by a capital "M" followed by the major outside diameter in mm and—for fine threads only—the thread pitch. A 2 mm pitch for coarse threads is understood, but not written.

The basic metric thread notation is explained in Fig. 6-9. Tolerance symbols consist of numerals and letters. The numerals define the tolerance permitted on both internal and external threads. The letters designate the position of the tolerances in relation to basic diameters. Lower case letters apply to external threads, whereas capital letters apply to internal threads. The numeral and letter combination is called the *tolerance symbol,* and it identifies the actual maximum and minimum limits for external and internal threads. The tolerance symbol can be applied to both the pitch diameter and crest or major diameter., as shown in Fig. 6-9.

SUMMARY

The SI system has three classes of units: base units, supplementary units, and derived units. At this point in your training, you need only be concerned with the base units. There are seven base units: length (meter), mass (kilogram), time (second), electric current (ampere) thermodynamic temperature (kelvin), amount of substance (mole), and luminous intensity (candela). However, blueprint reading requires mainly a knowledge of the unit of length, the meter and its sub-multiples: the decimeter (dm), centimeter (cm), millimeter (mm), and the micrometer (μm).

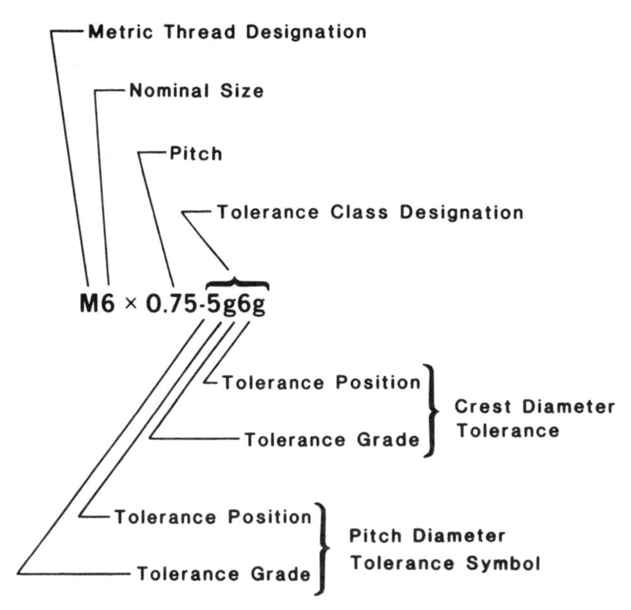

Figure 6-9 ISO metric thread notation.

Of these sub-multiples of the meter, the millimeter (mm) is used most on blueprints and engineering drawings. The abbreviation *mm* stands for the millimeter. Never abbreviate millimeter with capitals (*MM*), or you will change its meaning.

In the manufacturing environment, you will encounter three types of metric drawings: dual dimensioned, conversion chart metric, and all-metric. You can readily identify each type of drawing. The dual dimensioned drawing shows both the millimeter and inch units within the dimension line. Conversion chart drawings include a conversion chart to convert metric units to inch units, and *vice versa.* This type of drawing shows only one unit of measurement within the dimension lines. All-metric drawings give only the metric units on the drawing, and the drawing is usually stamped or marked "Metric."

Soft conversion applies to all dual dimensioned and conversion chart metric drawings, and to those all-metric drawings that do not conform to ISO standards. Soft converted drawings do not require special tooling.

Hard converted metric drawings conform to ISO standards, and are so indicated by a first-angle projection symbol. Hard converted drawings require special metric tooling for machining, as well as metric measuring instruments for inspection.

When converting from inches to millimeters and millimeters to inches, you must follow formal conversion guidelines. The reason for this is to maintain the accuracy of the original unit of measurement, and to insure interchangeability.

Metric screw thread notation is similar to the Unified Thread system, except that the metric system has only three thread series: Coarse, Fine, and Extra-Fine. The metric screw thread system uses no number-size classification. Metric threads are designated by a capital *M* followed by the major outside diameter.

The metric unit for surface finish is the micrometer (μm) instead of the microinch of the inch system. You can convert customary surface finishes to metric surface finish designations relatively easily. However, be very careful in making the conversion.

You need a good, practical understanding of the metric system of measurement, because it is being used increasingly in industry, and will become even more prevalent within the near future.

TRAINING PRACTICE

1. What are the three classes of units in the metric system? ______________________

2. What are the four important sub-multiples of the meter that you will encounter in manufacturing? ______________________

3. Of the four sub-multiples, which one is used consistently on drawings? ______________________

4. What are the abbreviations for millimeter and micrometer? ______________________

5. How many millimeters are in one inch? ______________________

6. How many millimeters are in one meter? ______________________

7. What is the correct way to write "one thousand millimeters"? ______________________

8. What are the three types of metric drawings you will encounter in industry? ______________________

9. What is the difference between hard and soft converted metric drawings? ______________________

10. Show the four methods of writing dual dimensions:
    a. ______________________
    b. ______________________
    c. ______________________
    d. ______________________

11. Which method is recommended? ______________________

12. How are first-angle projection drawings identified? Illustrate.

13. Why should you use caution and great care when converting from metric to inch units and vice versa? ______________________________

____

14. What symbol is used to indicate surface finish on metric prints? ______________________________

____

15. Which unit is used to measure surface quality? ______________________________

____

16. Is the root diameter of metric threads larger or smaller than Unified threads? ______________________________

____

17. How are ISO metric threads designated? ______________________________

____

Training Exercise #1

Refer to: TP-160

Answer the following questions:

1. Is this part drawn in first- or third-angle projection? ______________________

2. What material is the brake made from? ______________________

3. With fillets and radii showing in the corners of this part, what manufacturing process was used to fabricate the rough part? ______________________

4. Which type of metric drawing is this? ______________________

5. Is this a hard or soft converted drawing? ______________________

6. What measuring system does this drawing use? ______________________

7. How many millimeters are contained in 1 in.? ______________________
   One millimeter equals how many inches? ______________________

8. 0.01mm is equal to how many inches? ______________________

9. What alignment is required on the $15.862\ ^{+0.007}_{-0.000}$ dia holes? ______________________

10. What is the width of the rib containing the 15.862 dia holes? ______________________

11. What information on the print can you use to convert all dimensions to English measurement?
    ____

12. What does a 0.8√ surface finish correspond to in the inch units of measurement?
    ____

Training Exercise #2

Refer to: TP-161

1. Which dimensioning system is used in this drawing? ______________________________

____

2. How can you tell which values are metric and which are English when reading the dimensions contained within the dimension lines? ______________________________

____

3. Referring to the $\frac{2.4}{.093}$ CENTRAL dimension in the end view, which value is metric and which value is English? ______________________________

____

4. What size tap will be used to cut the left- and right-hand internal thread? ______________

5. Are these taps U.S. or metric? ______________________________

6. When machining this part, what would be a good practice to follow to make sure that you are working with the correct unit of measurement? ______________________________

____

____

7. What angle of projection is used on this print? ______________________________

____

____

Training Exercise #3

Refer to: TP-162

1. What type of metric conversion is used on this drawing, hard or soft? ____________________

2. Can a standard U.S. drill size be used to drill the 4.763 dia and 1.57 dia holes? ____________________

3. What is the surface finish required for the three radial grooves? ____________________

4. Is angular measurement (the way in which an angle is measured) different when using the metric unit of measurement? ____________________

5. Can this part be produced on conventional U.S. machine tools? ____________________
   Why? ____________________

6. What is the general tolerance for dimensions not showing specific tolerance? ____________________

7. What type of metric drawing is this? ____________________

Training Exercise #4

Refer to: TP-163

1. What is the name of the part? ______

2. To what scale is the part drawn? ______

3. What kind of finish will this part receive? ______

4. What material is the part made from? ______

5. Where did you find the information to answer Questions 1 to 4? ______

6. When was the last revision made to the drawing? ______

7. What was the nature of the revision? ______

8. What hardening process is to be used? ______

9. How hard will the part be when completed? ______

10. Name the views:
    a. ______
    b. ______
    c. ______

11. How is the alignment of the .3725 dia and .3405 dia to be checked? ______

12. 0.01mm is equivalent to how many tenths (ten thousandths of an inch)? ______

13. How are metric units and English units shown in each dimension? ______

14. What is the center-to-center distance (in mm) from the 13.72mm dia to the 2.03mm dia? ______

15. What is the distance (in mm) from Ⓐ to Ⓑ? ______________________

16. What is the maximum size of the [1.60mm] hole? ______________________

17. What general tolerance does a three-place decimal inch dimension have? ______________________

A two-place decimal? ______________________

18. How thick is the material? ______________________

19. What is the off-set distance of the vertical ℄ for the [2.03] .080 dia hole? ______________________

20. Are the .125 dia holes a functional feature of the completed part? ______________________

21. Are the two .125 dia holes centered in the trough (equidistant from the inner sides) of the part?

____

22. What is the overall length (English and metric) of the part? ______________________

23. What is the overall width of the right view? ______________________

24. Does the 1.60mm hole go through the entire width of the part? ______________________

25. What is the vertical center distance between the 1.60mm and 1.85mm holes? ______________________

Chapter VII

CASTINGS, FORGINGS, AND WELDMENTS

Machine shops use four main processes to produce most parts: casting, forging, machining from standard stock, or welding standard shapes. Each of these different methods produces parts that have a characteristic shape and appearance. Therefore, each method requires different details on engineering drawings.

CASTINGS

Several different casting processes have been developed to fill specific manufacturing needs, each having its own advantages and disadvantages. However, their similarity is that each process employs a mold cavity, which determines the desired shape and size of the casting. The mold may be filled either by gravity, or by forcing the required amount of liquid metal into the mold under pressure.

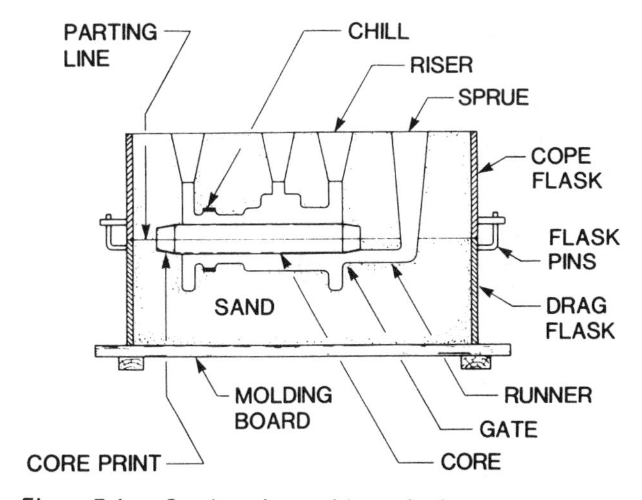

Figure 7-1 Sand casting mold terminology.

Sand Casting

Sand casting consists of making a cavity in a special sand with a pattern, and filling the cavity with molten metal. When the metal cools and solidifies, the sand is broken away. The casting is removed, trimmed, and cleaned. All molten ferrous and nonferrous metals can be sand-cast.

Sand molds are made in two sections, as shown in Fig. 7-1. The top section is called the *cope,* and the bottom section is called the *drag.* The *flask* is the frame that holds the sand. The joint between the sections is called the *parting line.* Molten metal is poured into the hole, called the *sprue,* and flows down *runners* to the *gate,* which is a passageway cut to the cavity. The metal flows by means of gravity. The *riser* or *feedhead* is a reservoir of molten metal that supplies additional metal as the casting solidifies and shrinks. Each type of metal shrinks a different amount, depending upon its composition. All sand-casting molds are destroyed as the solidified casting is removed from the mold cavity.

The pattern used in sand molding is a wood or metal duplicate of the product to be made, with some modifications. For example, it is made slightly larger than the final product, to compensate for the metal shrinkage during cooling. This is called *shrinkage allowance.* Patterns are also modified by a slight taper called a *draft,* which is applied to surfaces parallel to the direction of withdrawal of the pattern. The draft makes it easier to remove the pattern from the mold, and lessens the tendency of the mold walls to break away.

Another modification to the pattern is the *machining allowance.* This is the excess stock added to the pattern so that the casting will be oversize at those surfaces that require finishing by subsequent machining operations. Finishing is necessary at critical locations on products where mating surfaces contact one another, or when the as-cast dimensional precision is insufficient.

Large solid castings are expensive, and costs are reduced whenever metal can be saved. Therefore, cores are used extensively to reduce the amount of metal needed. Cores are also used in molds wherever a hole or undercut must be produced in a casting. The *core* consists of a rigid mixture of synthetic sand bonded by a special core oil or proprietary binders. *Core prints* position and hold the core in the mold cavity. They are extensions on the pattern that form a recess at the outer edges of the mold cavity into which identical extensions of the core are fitted. The core is inserted into position manually after the pattern has been withdrawn from the sand mold. It is destroyed as it is removed after the casting has cooled and solidified.

Precision Investment or Lost-Wax Casting

Investment casting is used to produce parts having complex shapes from alloys with very high melting points. These metals are usually difficult to machine or shape using conventional machine tools. The process uses heat-disposable patterns surrounded by a shell of refractory material to form the mold. Castings form in the cavities, created by melting out the pattern.

In precision investment casting, wax or plastic patterns are made first by pouring or injecting the pattern material into master die sets made to a high degree of accuracy by skilled diemakers. A cluster or "tree" is created by assembling several patterns together, gated to a central sprue by wax-welding. Next, solid patterns are made by placing a metal flask over the cluster, and then pouring a hard-setting molding material into the flask. The material completely invests the pattern cluster, and forms shell patterns. These clustered shell patterns are dipped repeatedly into a ceramic slurry, until the required thickness of the mold or shell is achieved. The flask and its contents are then placed in an oven, where the wax patterns are allowed to melt and drain out of the mold. Molten metal is then poured into the cavity, usually by means of gravity. Other pouring methods are also used, such as vacuum, centrifugal, and air pressure.

After cooling, the mold is broken away, freeing the castings. The cores are removed by a pressurized water blast, or by leaching with a caustic soda. The final operation is to separate the castings from the gating system with band saws or abrasive wheels.

Permanent Mold Casting

Permanent mold casting is simply the process of gravity-pouring molten metal into metal molds. It is also called *gravity die casting.* This process is especially suited to high-volume production of small, simple castings that are reasonably uniform in wall thickness, have no undercuts, and have limited coring requirements. It can also be adapted to make moderately complex castings.

The metal molds are made in two or more sections, hinged and clamped for easy removal of the solidified casting. Both metal and sand cores are used to form the cavities in cast parts. (When sand cores are used, the process is called *semi-permanent mold casting.*) To produce molds having a superior surface finish, close dimensional tolerance, and proper draft angles, the mold cavity is first formed to its approximate shape by a casting process, and then machined to final size. Multiple-cavity molds can be made.

Centrifugal Casting

Centrifugal casting is used to improve the mechanical properties of the metal casting. There are three principal methods of producing centrifugal castings: true centrifugal, semicentrifugal, and centrifuge. Each produces parts having particular shapes. In each method, molten metal is poured into a rotating mold. Centrifugal force causes the metal to be thrown and held against the mold wall until it solidifies.

Die Casting

Die casting is a fast and economical process for producing parts that do not require great strength. First, molten metal is forced into the cavities of metal dies under pressure, and held for a short time until the metal solidifies. The die blocks are then opened, and the casting, with its assembly of sprue, runners, and gates, is ejected by pins. The die is closed, and the cycle is repeated. Although die casting employs permanent molds or dies, do not confuse it with permanent mold casting, wherein the mold cavities are fed entirely by gravity force.

The alloy steel dies used in die casting usually consist of two sections that meet at a vertical parting line for casting removal. One half is called the *cover half*, and the other is called the *ejector half*. Multiple cavities can be machined into the two mating faces of the die block. The die block is held in a die casting machine, locked securely by clamps or toggle linkages. To hasten cooling and to prevent the formation of air pockets in the casting, the die is often water- or air-cooled and vented.

Detail Drawings for Castings

The pattern shop usually receives the same drawing to work from as the machine shop, showing all dimensions and finish marks. The metal to be used and generally the pattern number will be stated in the title block. The patternmaker provides for machining allowances, using the finish marks as his guide. Because metals shrink as they cool, the pattern must be made slightly oversize. Thus, the patternmaker uses a special *shrink rule* whose units are correlated with the shrinkage characteristics of the metal to be used, approximately 1/8 in. per foot. The patternmaker also provides the draft, or slight taper, which is not usually shown on the drawing.

Blueprints of castings are easy to identify. They show many fillets, rounds, bosses, and webs because castings, unlike forged or rolled metals, are quite brittle. They chip or break easily.

FORGINGS

Forging is a process that plastically deforms metals to a specific shape by means of a compressive force exerted by a hammer, a press, rolls, or an upsetting machine. Forging actually compresses the molecular structure of the metal as it conforms to the predetermined shape, thereby making the part stronger than a casting, which does not have a similar molecular pattern. Top-quality tools (wrenches, hammers, etc.) are forged so that they can withstand repeated shock and loading forces. In addition, intricate shapes can be forged to exacting specifications. Therefore, the characteristics of precision locations and sizes, strength, and intricate shape can be obtained in a single part through accurate forging practices. In conventional forging, the metals are heated to improve the plasticity of the metal, and to reduce the forging forces required. However, some special forging processes employ cold-forming techniques.

Forged parts can be produced in either open or closed dies. *Open-die forging* progressively applies local compressive forces to different parts of the metal stock. Parts produced by a *closed-die forging* process are formed by applying force to the entire surface, causing the metal to flow into a die cavity cut to a specific shape. More complex and accurate parts are produced with closed dies than with open dies.

Smith Forging

Smith forging (also known as *flat-die forging*) is an open-die process that shapes a heated metal part by applying repeated blows of a hand-held hammer on a flat die or anvil. The smith maintains the desired length and cross section of the part by manually adjusting the position of the part on the flat surface of the anvil. Only relatively simple shapes can be forged with this method.

Hammer Forging

Hammer forging differs from smith forging in that the heated billet is shaped by the impact of a steam- or air-operated hammer instead of manually. This makes it possible to forge larger heavier parts, because higher pressures can be exerted on the workpiece. The horizontal faces of the hammer (upper die) and the anvil (lower die) are both flat. Mechanical manipulators hold and position the work that is too heavy to be positioned manually. Special tools, attached to long safety handles, allow the hammer man to produce holes with punches and specially-shaped dies, perform notching or cutoff operations with chisels, and grip the parts while forging special configurations. Like smith forging, parts produced by hammer forging are restricted to rather simple shapes.

Drop Forging

Drop forging forms metal parts by hammering a heated bar or billet into aligned die cavities. Either steam- or gravity-hammers are used. The dies are made in sets or halves. One half is attached to the hammer, and the other to a stationary anvil. Parts are not formed by a single blow of the hammer, but are progressively developed in a sequence of closed-impression die cavities of successively different shapes. Each impression gradually distributes the flow of metal, and changes the shape of the workpiece as it transfers from one impression to the next between hammer blows.

Press Forging

In press forging, parts are made by large vertical presses, which plastically deform a blank into one or more die cavities using a slow squeezing action. The presses are either mechanically- or hydraulically-operated. The parts are formed by a single stroke in closed-impression die sets. Completed forgings are manually removed from the die cavity. Most press-forged parts are first preformed on other machines, such as forging rolls, upsetters, benders, or special machines. The press is used only for the blocking and final forging operations.

Upset Forging

Upset forgings, also known as *machine forgings,* are produced in double-acting mechanical presses that operate in a horizontal plane. Upsetting is performed by the impact blow of a punch against the end of a piece of stock inserted into a shaped die. The press action squeezes the desired shape of the heated metal workpiece into the die cavities.

Roll Forging

Roll forging reduces the cross section of short lengths of barstock, while increasing the stock length. Pairs of semi-cylindrical rolls are used, with the active die surface usually occupying only half of the roll circumference. First, the stock is manually inserted between the roll dies. Shape formation begins as the revolving rolls grip the stock, squeeze it, and roll it back toward the operator. As the first pass is completed, the rolls open and the operator removes the bar and inserts it between another set of grooves for the next pass. The procedure is repeated, using progressively smaller die impressions, until the workpiece reaches the desired size and shape. Both tapered and straight work can be roll-forged.

Close Tolerance Forging

In close tolerance forgings, draft angles are on the order of 1 to 3 degrees, and tolerances are less than half those of commercial forgings. Close tolerance forgings have little or no allowance for finish.

Precision Forging

Precision forgings have close dimensional tolerances requiring special tooling and equipment. Precision forgings are usually made from aluminum and magnesium.

Detail Drawings for Forgings

Forged parts look somewhat like castings, because they use many fillets and rounds to improve strength. However, unlike brittle castings, forgings are tough and resilient. A typical forging drawing will show the unmachined forging and the machined forging on one sheet of paper. The drawing of the unmachined forging carries the dimensions that the diemaker needs to make the forging dies. Because draft must be provided to release a forging from the forging dies, it is shown on the drawing and dimensioned, usually by a note. The dimensions parallel to the horizontal surfaces of the die are also usually given to specify the size at the bottom of the die cavity. The parting line of the die set is also shown.

MACHINED PARTS FROM STANDARD STOCK

The shape of a part often lends itself to machining directly from standard stock material, such

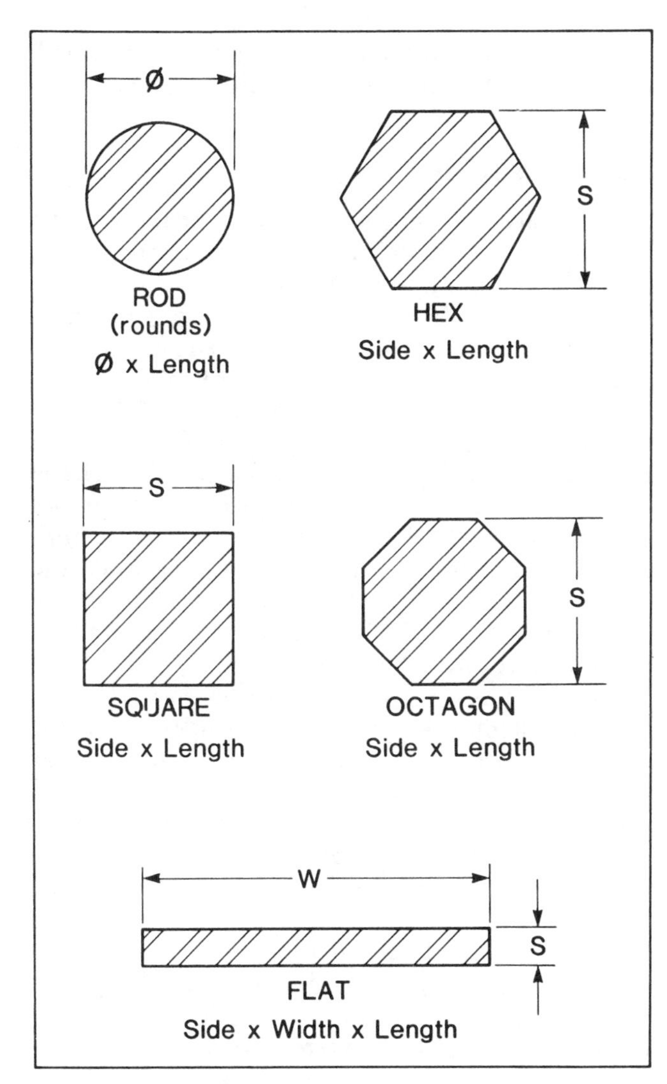

Figure 7-2 Bar

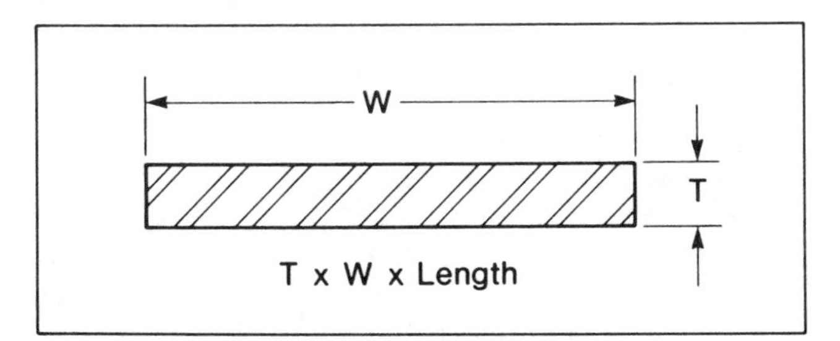

Figure 7-3 Plate

as rods, bars, plate, tubing, and standard structural shapes. There are many different shapes of steel and aluminum available in stock form. It is important that you know these shapes and how they

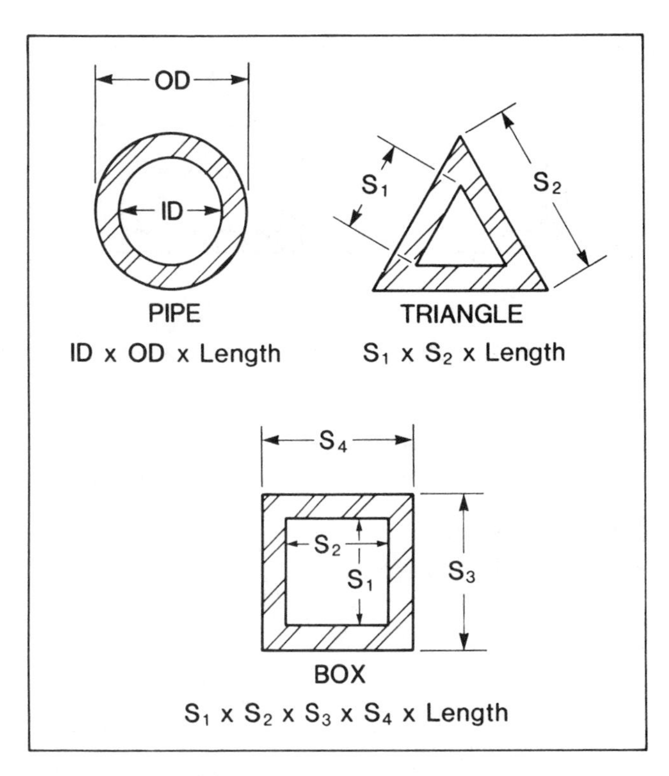

Figure 7-4 Tubing

are measured. Both the size and the grade are specified on drawings. Parts produced from stock are usually finished on all surfaces, and the general note "Finish all over" on the drawing eliminates the need for finish marks. All functional dimensions are provided.

The illustrations of standard stock shapes in Figs. 7-2 to 7-5 will be helpful to you when you must order stock or make up a bill of materials for a job.

WELDMENTS

Because castings are costly in terms of personnel, equipment, and space, and welding technology has improved substantially over the years, weldments or fabricated welded parts have increasingly replaced castings. Weldments are quick to prepare and cost less to repair, whereas castings require a foundry, foundry workers, and patternmakers, plus large amounts of time to make patterns and molds, pour the iron, and cool and age the part. Today, weldments can be constructed on-site by one or more welders. This represents a major advance in manufacturing flexibility, in that the skills needed to build equipment are the same ones needed to

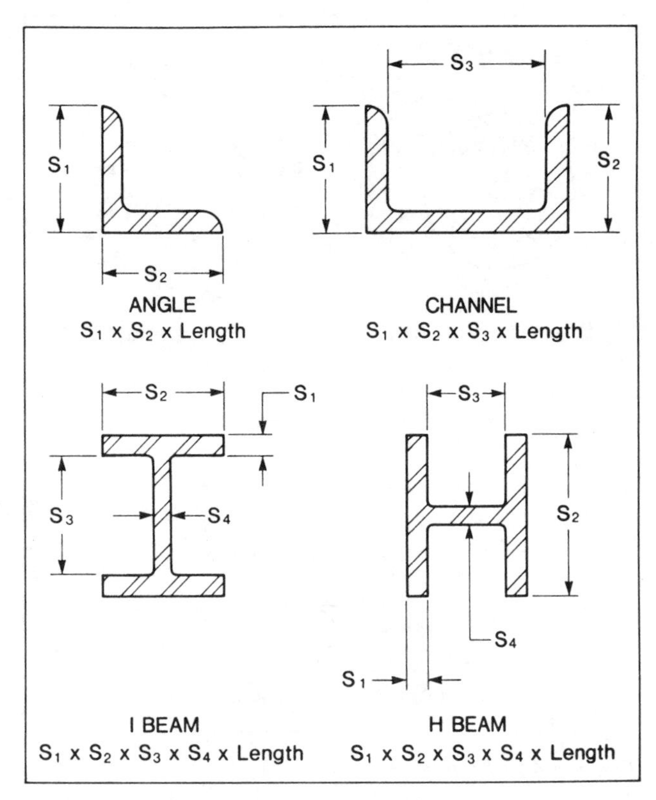

Figure 7-5 Structural.

repair and modify it. The only post-construction requirement for a weldment is stress-relieving, and sometimes shot-blasting.

Arc and gas welding consist of permanently fastening together two or more pieces of metal into a single homogeneous part by applying heat.

Arc Welding

Arc welding is the major welding process used in industry, largely because of its speed. In this process, intense heat is generated by an electric arc struck between a consumable welding electrode (or rod) and the workpiece. The weld metal both melts and solidifies very quickly.

Gas Welding

The most common form of gas welding is *oxy-acetylene welding,* wherein commercially-pure acetylene gas and oxygen are combined in proper proportion to produce a high-temperature concentrated flame. This flame fuses the melted metal without requiring pressure. Because normally a slight gap exists between the pieces being joined, filler material in the form of a *wire* or *rod* is added during the process.

Weld Symbols

An enlarged drawing of the standard welding symbol is shown in Fig. 7-6, with explanations of where the marks and size dimensions that completely describe a weld are found on a drawing. The arrow is the most important element of the symbol. It always points to the joint to be welded. If the weld is on the *arrow side* (see Fig. 7-7A) the symbol showing the type of weld (fillet weld in Fig. 7-6) is placed below or to the right of the baseline attached to the arrow. The weld type symbol will be below the baseline when the line is horizontal, and to the right of the baseline when the line is vertical. When the weld is located on the *other side* (see Fig. 7-7B), the symbol will be above or to the left of the baseline.

To indicate that a weld is to be on *both sides,* the same symbol is placed on both sides of the line (see Fig. 7-6 and 7-7C). If the weld is to be made *all around* (as when a piece of tubing is welded to a plate), a circle is drawn around the joint between the baseline and the arrow, as shown in Fig. 7-6 and 7-7D. When the weld is to be made in the *field,* a solid dot is also placed at the junction between the baseline and the arrow, as shown in Figs. 7-6 and 7-7E. A typical drawing for gas or arc welding of a part is shown in Fig. 7-8. The size of the weld is given next to the weld symbol, as well as the length and the pitch. Weld dimensions are given in inches. When a weld is discontinuous (consisting of a series of like welds or increments), as in Fig. 7-8, the length of the increment, and the length between the increments, are given.

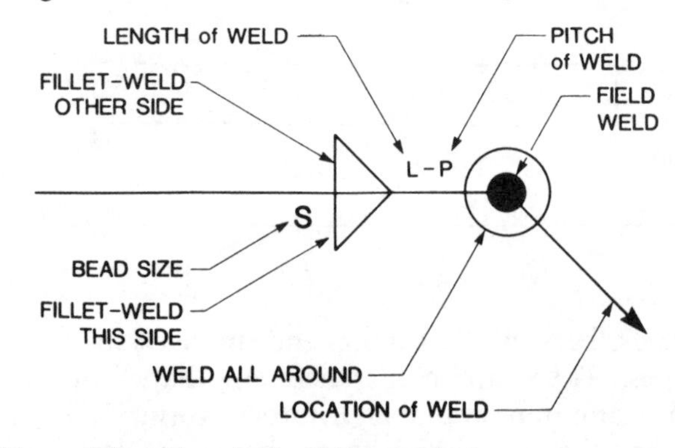

Figure 7-6 Standard welding symbol.

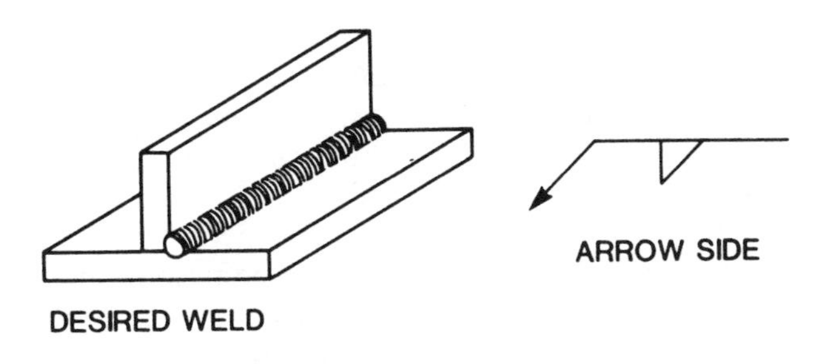

Figure 7-7A Arrow side

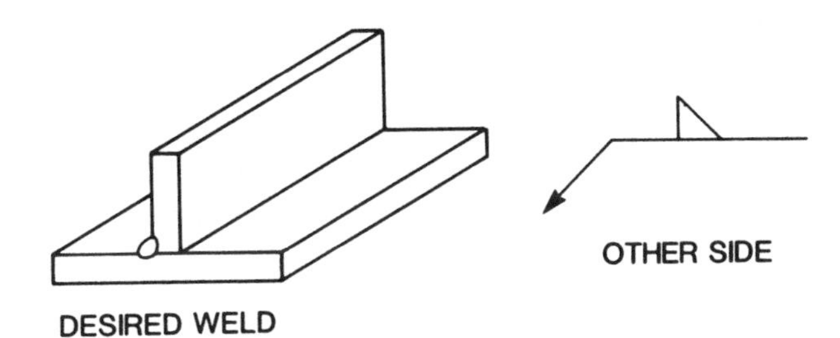

Figure 7-7B Other side

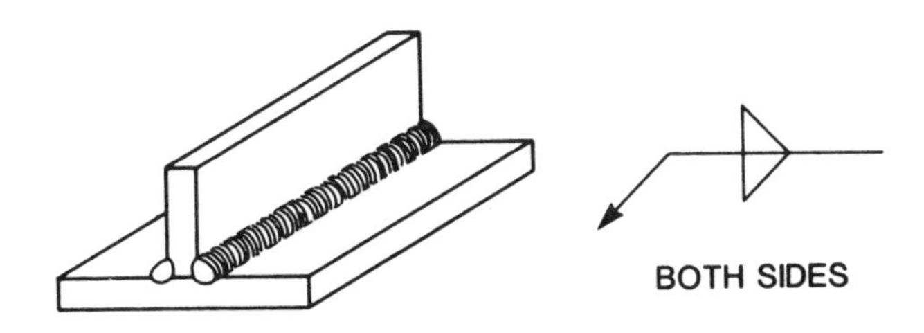

Figure 7-7C Both sides

Figure 7-7D All around

Figure 7-7E Field

Weld Joints

To insure the maximum strength that a weld is capable of, weld joints are often prepared (or "prepped") before welding with specially-machined grooves, such as those shown in Figs. 7-9 to 7-14.

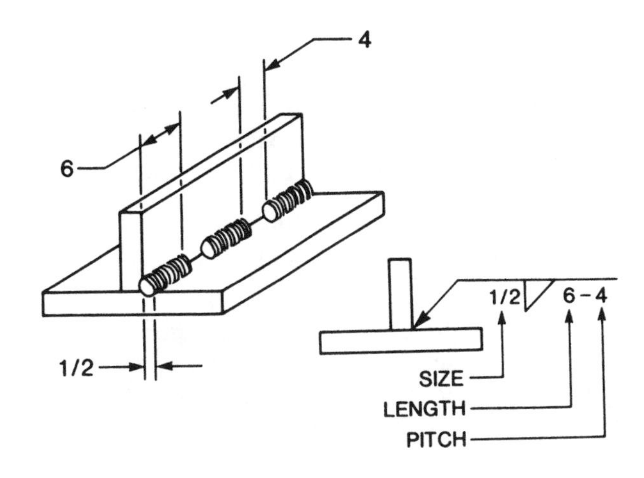

Figure 7-8 Typical drawing for gas or arc welding.

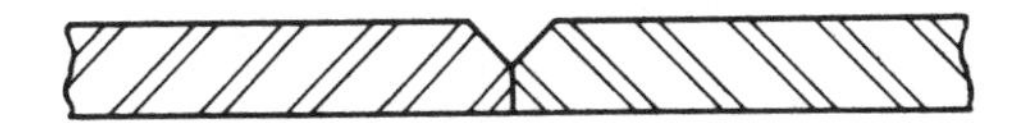

Figure 7-9 "V" groove—most basic prep. Resembles an oversized chamfer. Sizes are identified the same way that chamfers are.

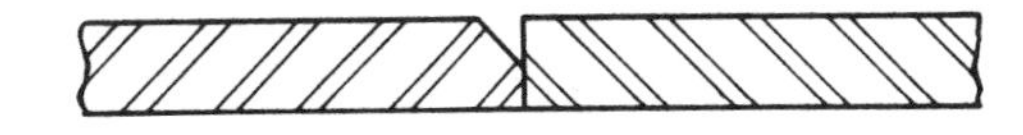

Figure 7-10 Bevel groove—measured as "V."

Figure 7-11 "J" groove—popular prep shape for high-stress welds. Sizes are determined by radius and depth.

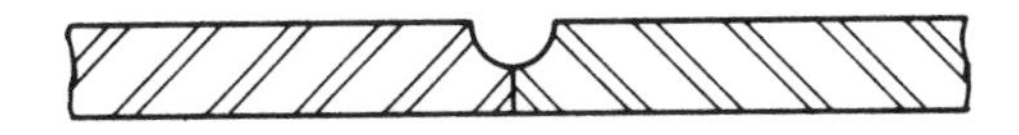

Figure 7-12 "U" groove—composed of two adjacent "J" grooves.

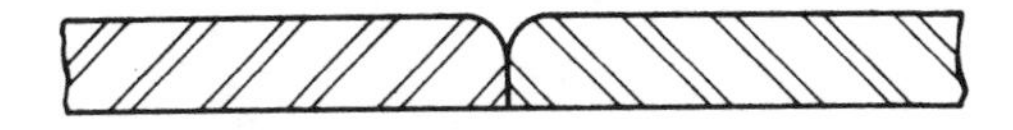

Figure 7-13 Flare—"V" groove.

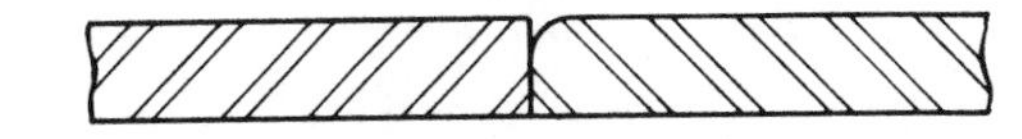

Figure 7-14 Flare bevel groove.

SUMMARY

In addition to machining simple solid parts from standard stock shapes, parts are produced for finish machining operations by three process:

(1) Casting
(2) Forging
(3) Welding

Each of these manufacturing processes has certain advantages of strength, cost, or flexibility. Castings are inexpensive and can be intricate in shape—more so than forged parts where large volume production is required. Forged parts are strong and durable. Weldments are inexpensive and easy to construct where short run items or very large parts are to be produced.

You can identify castings on blueprints by the function of the part (no heavy loading nor shock conditions evident), and the wide use of webs, fillets, rounds, and bosses. Forgings are also recognizable by part use and the application of many fillets and rounds. You can recognize weldments through the use of welding symbols and structural metal shapes in construction.

TRAINING PRACTICE

1. Given the type of part or its use below, specify the basic manufacturing process used to make it (casting, weldment, or forging).
   a. Railroad car wheels ______
   b. Major carburetor components ______
   c. Large punch press frames ______
   d. Wrenches ______
   e. Rings (jewelry) ______
   f. Machine tool frames ______

2. Name the characteristics of each process that are apparent on a blueprint.
   a. Weldments ______
   b. Castings ______
   c. Forgings ______

3. List all the information necessary to order the following structural shapes:
   a. Pipe—3 in. OD, 2 in. ID, and 15 ft long. ______
   b. Angle—1 in. x $1\frac{1}{2}$ in. sides, $\frac{3}{16}$ in. thick, and 18 in. long. ______
   c. Plate—3 ft wide, 6 ft long, and 3 in. thick. ______
   d. Hexagon bar stock—2 in. over flats, $2\frac{1}{4}$ in. over points, and 28 in. long. ______

Chapter VIII

SKETCHING TECHNIQUES

Most machinists keep a pad of paper, pencil, and rule (or 12 in. scale) in their toolboxes to make a sketch from an existing part or mechanism, or from an assembly or detail drawing. One helpful practice that apprentices are encouraged to use is to take notes on a particular job. Just make simple sketches of the work and setup, adding a few brief words of description. Because complex jobs may recur, a quick look at the sketches and notes made earlier can save you a lot of time. In addition, you won't have to remember how you did a successful job previously.

In the planning stage of developing a jig or fixture, tool and diemakers use freehand sketches to organize the sequence of operations, and to record information on standard parts. These sketches are a great help in preparing the accurate design layouts needed, particularly for complex jobs and where different craftsmen will be working on the same job. As you learned in Chapter I, nothing communicates an idea faster and more accurately than a drawing.

The single most important thing to remember in freehand sketching is to be consistent in all your drawings. Consistent means that when you begin to sketch, follow all the rules and conventions of blueprint construction. Make sure that six months after you have made a sketch, you will be able to read and understand it completely.

SKETCHING MATERIALS

Freehand sketching, sometimes called *freehand drawing,* is so named because you use no drafting instruments, such as a T-square, straight edges, french curves, or compasses. One of the advantages of freehand sketching is that it requires only paper and pencil—items that anyone has for ready use. However, the type of sketch you make, and your personal preference will determine the materials you use.

A soft-lead pencil is best for sketching, somewhere in the grade range from F to 3H, with H being a good grade for most sketching. The pencil should be long enough to permit a relaxed, but stable, grip. As you gain experience, you may prefer to use fine-tip felt pens. (Dark- or bright-colored pens work best.) Felt-tip pens work very well on overlay sketches, described in the next paragraph.

You will draw most of your sketches on scratch paper, which can be any type or size. A 3 in. x 5 in. or 5 in. x 8 in. scratch pad is handy to carry in a toolbox. Tracing paper is convenient to carry, too, for planning the layout of a drawing. The advantage of sketching on tracing paper is that you can change or redevelop sketches easily, simply by placing transparent paper over (overlaying) your previous sketches or existing drawings. Sketches prepared in this manner are called *overlay sketches.* You can also use cross section or graph paper (see Fig. 8-1) to save time if you must draw sketches to scale. Isometric sketches are easily made on specially-ruled isometric paper, as shown in Fig. 8-2. You may want to carry a pink rubber eraser too, although you will probably do very little erasing of final lines. You can usually redraw sketches faster than you can erase them. To make

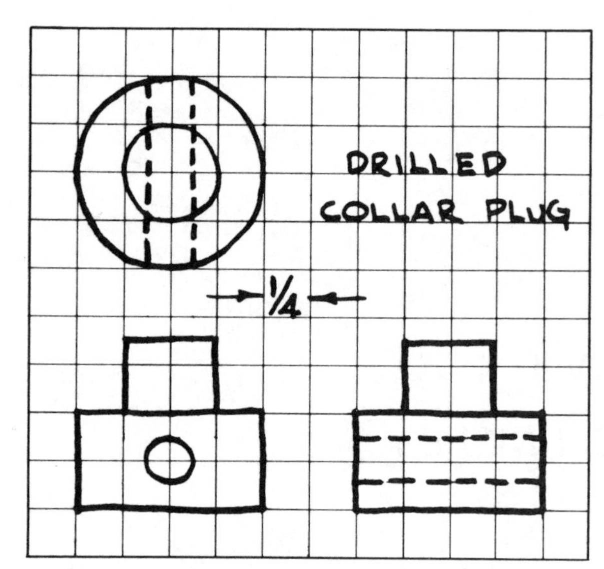

Figure 8-1 Cross section paper.

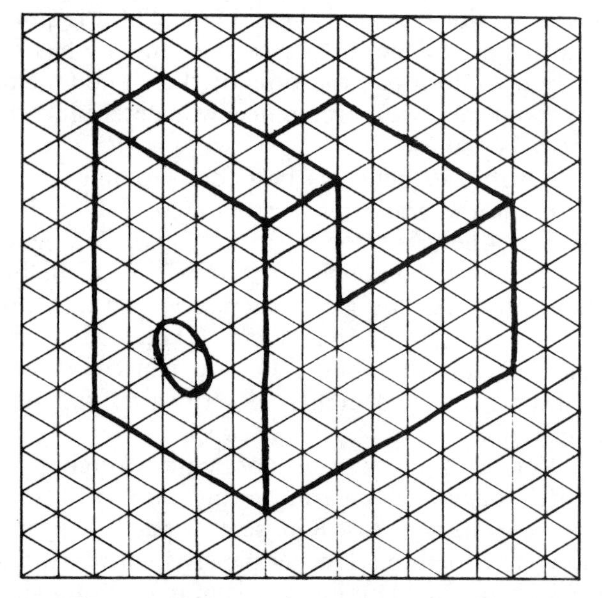

Figure 8-2 Ruled isometric paper.

dimensioned sketches, you will want to carry some sort of measuring instrument, such as a rule or metal tape.

SKETCHING LINES AND CURVES

All sketches consist of straight and curved lines. The main difference between freehand sketches and mechanical drawing is in the techniques used to make the lines and curves on the paper.

Straight Lines

No one can draw a perfectly straight long line, but anyone can draw a straight short one—about 1½ in. long. To draw a long line, place a dot at the beginning and end, and several in between. Then, simply connect the dots using a series of short pencil strokes, instead of one long one. By using short strokes, you can better control the direction of the line and the pressure of the pencil on the paper. Hold the pencil about ¾ in. to 1 in. from its point so that you can see what you are doing. Try to make free and easy back-and-forth movements, rather than tightening up your finger and wrist muscles into a cramped position. In other words, *relax.*

To keep your sketch neat, draw your lines lightly first. Lines not essential to the drawing can be sketched so lightly that you need not erase them. Later, when you have drawn what you wanted to, darken the essential lines by running the pencil over them using more pressure.

Drawing horizontal lines is easier for many people than drawing vertical or diagonal lines. If this is the case with you, draw all horizontal lines first, then rotate the paper 90 deg or less, as required, and draw the vertical and diagonal lines as if they were horizontal.

Dividing Lines and Areas Equally

You must be able to divide (bisect) lines and areas into equal parts to arrive at many of the common geometric figures that make up parts. The easiest way to bisect lines is by visual comparison, as shown in Fig. 8-3. Simply "eyeball" the entire line, and determine its center by optically comparing the two halves. You can repeat this pro-

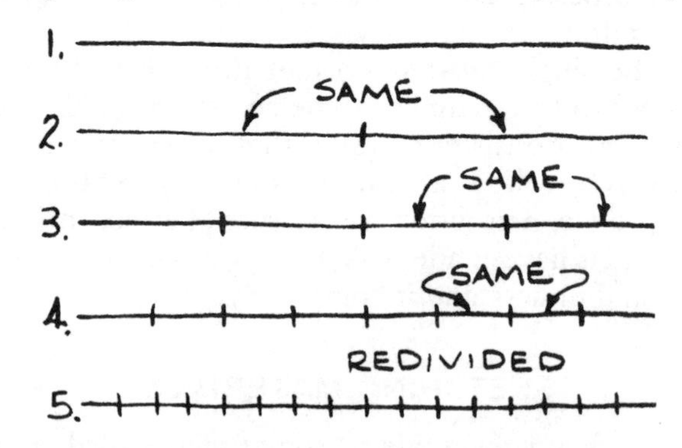

Figure 8-3 Bisecting a line by visual comparison.

cedure any number of times to divide a line into any number of equal parts, merely by dividing and redividing the line segments.

You can easily determine the centers of rectangular areas by first drawing their diagonals. If necessary, you can also divide the halves with diagonals to obtain smaller divisions, as shown in Fig. 8-4.

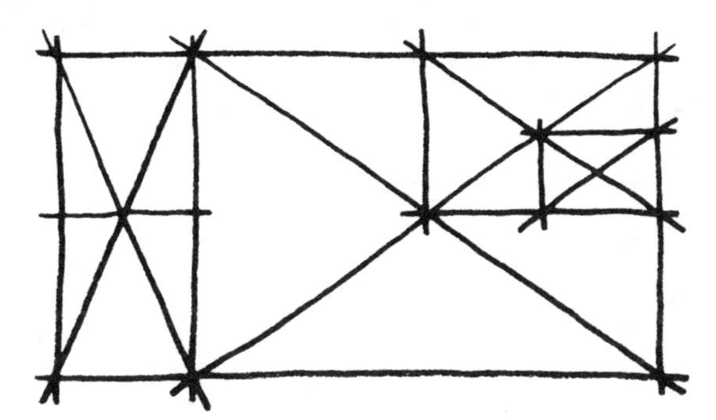

Figure 8-4 Finding the centers of rectangles by drawing diagonals.

Sketching Angles

You will require the 90 deg angle in many of your sketches. Use the perpendicular edges of your paper to serve as a visual guide when drawing them. Check their accuracy by turning your sketch upside down. This makes evident any non-perpendicular tendencies of horizontal and vertical lines. Check your 90 deg angles with a triangle occasionally to make sure you are drawing them correctly.

You can make a 45 deg angle by dividing a right angle using visual comparison, and you can make a 30 deg or 60 deg angle by dividing the right angle into equal parts in the same way (see Fig. 8-5). By starting with a right angle, you can make the most accurate estimate of the shape of any angle that divides into 90 deg equally.

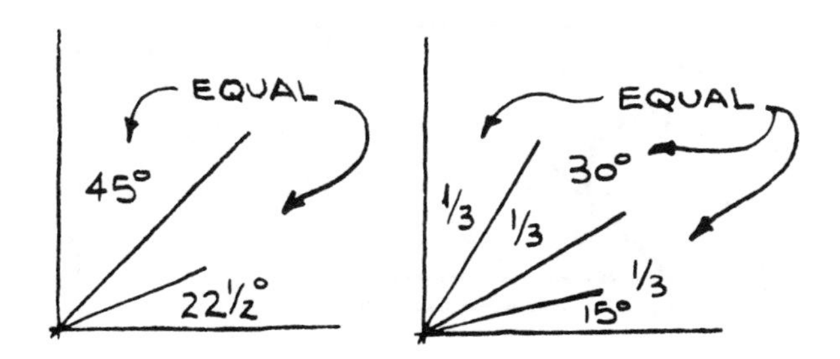

Figure 8-5 Sketching angles by visual comparison.

Sketching Circles and Arcs

Perfectly round circles are the most difficult to draw freehand. However, Fig. 8-6 shows how to draw both circles and arcs, using straight lines as construction lines. First, draw two straight lines crossing each other at right angles as in Fig. 8-6A. Their intersection is the center of the circle. Mark a piece of scrap paper to measure the radius of the circle from the center on each axis. Next, sketch a square with the center of each side passing through the marks on the axes, as shown in Fig. 8-6B. Now sketch in the circle, using the angles of the square as a guide for each arc.

When larger circles are required, you can add 45 deg angles to the square to form an octagon.

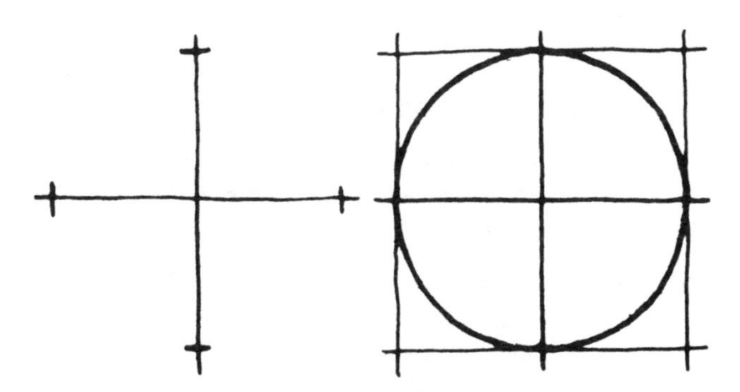

Figure 8-6 Sketching circles using construction lines.

This will give you four more points of tangency for the inscribed circle. Or, to draw very large circles, you can make a substitute compass with a pencil, a length of string and a thumbtack. Tie one end of the string to the pencil near the tip. Measure the radius of the circle you are drawing on the string, and insert the tack through the string at this point and into a pad of paper. Now, swing the pencil in a circle, marking the paper at the same time, and keeping the pencil perpendicular to the paper.

Sketching Curves

You can sketch curves by blocking them with straight lines, as shown in Fig. 8-7. This method is simply a variation of Fig. 8-6, which shows how to sketch circles. One of the best ways to sketch curves connected to straight lines is the six-step method illustrated in Fig. 8-8:

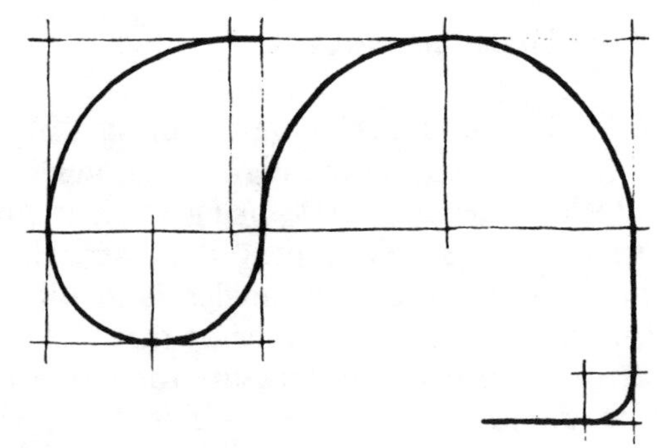

Figure 8-7 Sketching curves.

(1) Intersect a vertical and a horizontal line lightly, as shown in Fig. 8-8A.

(2) From the intersection, mark off the same distance on the vertical and horizontal lines, as in Fig. 8-8B.

(3) Draw the hypotenuse of the triangle lightly through the two points marked, as in Fig. 8-8C.

(4) Place an *x* or a dot in the exact center of the triangle formed, as in Fig. 8-8D. (Use visual comparison to find the center.)

(5) Start the curve from one point of the triangle (the intersection of the hypotenuse and the vertical line is best), as in Fig. 8-8E, touching the centerpoint and ending at the other point of the triangle.

(6) Erase all unnecessary guidelines, and darken the curve and any adjoining straight lines as shown in Fig. 8-8F.

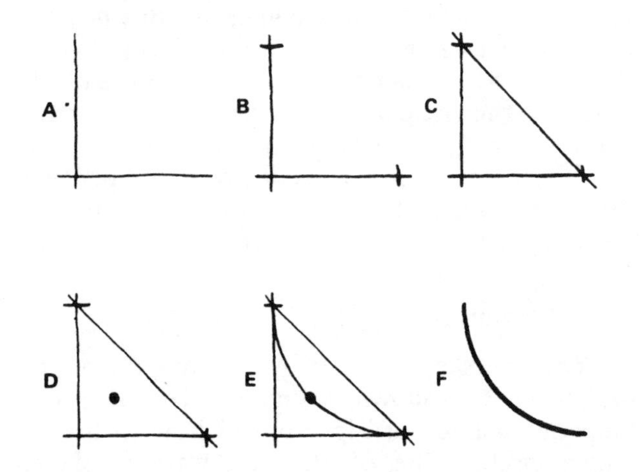

Figure 8-8 Sketching curves connected to straight lines.

Using Construction Lines

To sketch a part, such as the one shown in Fig. 8-9, you can't simply start at one corner and draw it, detail by detail, and have it come out with the various elements in correct proportion. You should first block in the rectangular outline of the object, as shown in Fig. 8-9A. Then draw light guidelines at the correct angles to obtain the shape of the object, as in Figs. 8-9B and C. Finish the sketch by drawing in the details, and then darken the outlines, as shown in Fig. 8-9D.

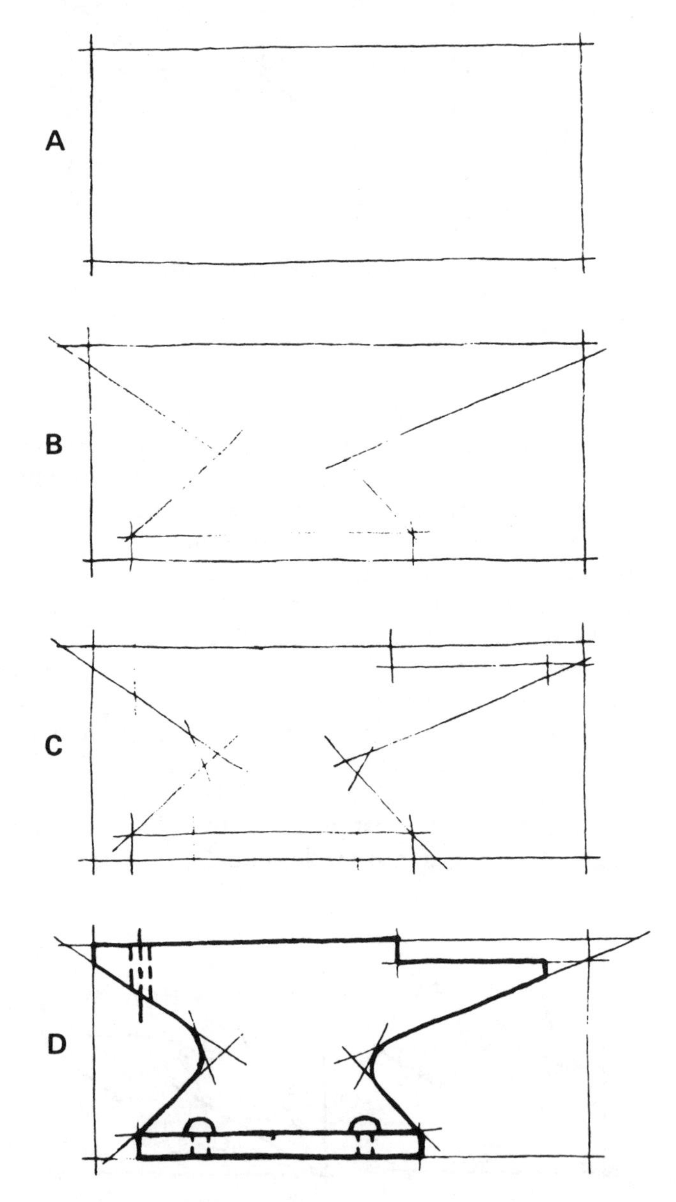

Figure 8-9 Using construction lines to sketch an object.

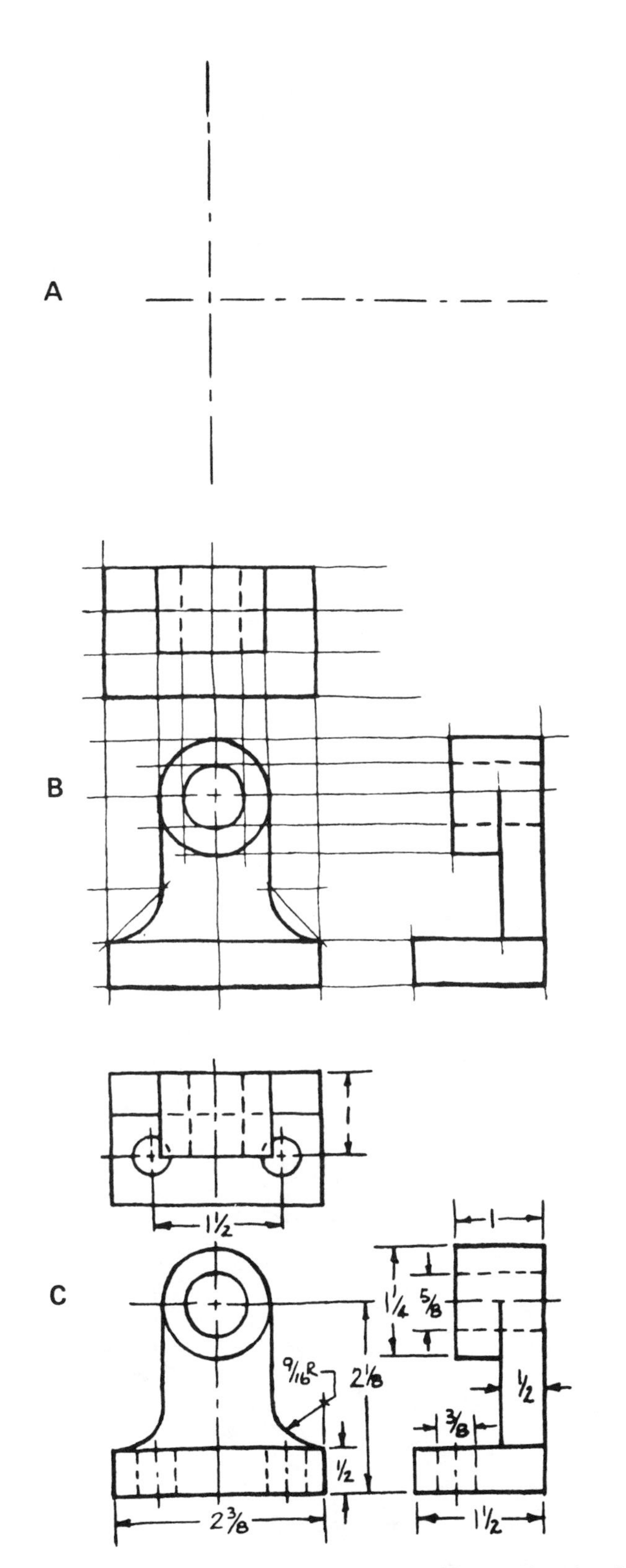

Figure 8-10 Sequence of an orthographic working sketch.

Line Widths

So that anyone can read your sketches easily and accurately, use the same relative width of lines to represent the same standard meanings as those described in Chapter II. To obtain the widths required, first sharpen a pencil to a conical point with a pencil sharpener. Then round it to the varying degrees of dullness required with a piece of sandpaper. Draftsmen use small pads of sandpaper made for that purpose. All final lines should be very dark, except construction lines, which should remain very light.

MAKING A WORKING SKETCH

To make a working sketch (a sketch that is dimensioned), start with a clean piece of paper, either plain, cross sectioned, or isometric. Estimate the size that the sketch should be, and select the view or views that will best represent the object —usually the front view. Then draw the orthographic projections or pictorial drawings, leaving enough space for dimensions. Follow this sequence of steps to make orthographic projections:

(1) Draw the centerlines as shown in Fig. 8-10A.

(2) Block in the views.

(3) Draw the outlines, aligning them as in Fig. 8-10B.

(4) Add the details to the surface of the views.

(5) Darken the lines of the finished sketch.

(6) Use an artgum or kneaded rubber eraser to erase the construction lines no longer needed. If necessary, touch up any lines you may have erased by mistake.

(7) Draw all required extension and dimension lines.

(8) Letter in the dimensions (see Fig. 8-10C).

You can see that the final working sketch provides all the information that you need to make the object.

Measuring Objects

Which measuring devices you will need when drawing sketches from existing structures depends upon the degree of accuracy *originally required* to produce the object, and the purpose of the sketch. If it is to be a freehand working drawing, it must carry all the dimensions needed to make the part. For sketches of ordinary machine parts, a steel scale and a set of outside and inside spring calipers are sufficient. You can often find the approximate radii of large arcs by lightly tracing the curve from the model, and then drawing it with a compass after you have determined the radius by trial and error. Sometimes you can transfer irregular shapes to the drawing paper by pressing the paper along the edges of the shape until they leave a dented impression. Then draw a line over the impression. To obtain very precise dimensions, you may need a surface plate and surface gage, or other precision measuring devices.

Dimensioning Rules

(1) Always make your sketch large enough to include all necessary dimensions without altering the sketch and making detail hard to read.

(2) Use the methods of in-line and base-line dimensioning discussed in Chapter IV. Don't place dimensions *on* the object. (Review the pages in Chapter IV on "Special Dimensions" for the proper methods of assigning sizes, positions, and notes.)

(3) Make extension line lengths consistent in size throughout the sketch.

(4) Include all necessary dimensions and notes on tolerancing, but don't over-dimension.

(5) Sketches that show both metric and English units of measure should be noted as such. The method you use to distinguish between the two must be consistent throughout.

Orthographic Sketching

In Chapter II, you learned that an orthographic drawing shows multiple views of the true shape of an object. Usually the three principal views (top, front, and right side) tell everything needed to make an object. Whether you are sketching from an object or just trying to communicate an idea, first define the different views of the object. If it's a small object, move it around on a flat surface to find the views that best show its true size and shape. If it's too heavy to move around or very large, move around *it*. In either case, choose the view that best shows the overall shape of the object for the front view. Then decide which other views will show the object in detail. Although two or three views will usually give a complete picture, more complex objects may require auxiliary views or sectional views.

Follow the sequence of steps under "Making a Working Sketch" to draw an orthographic sketch.

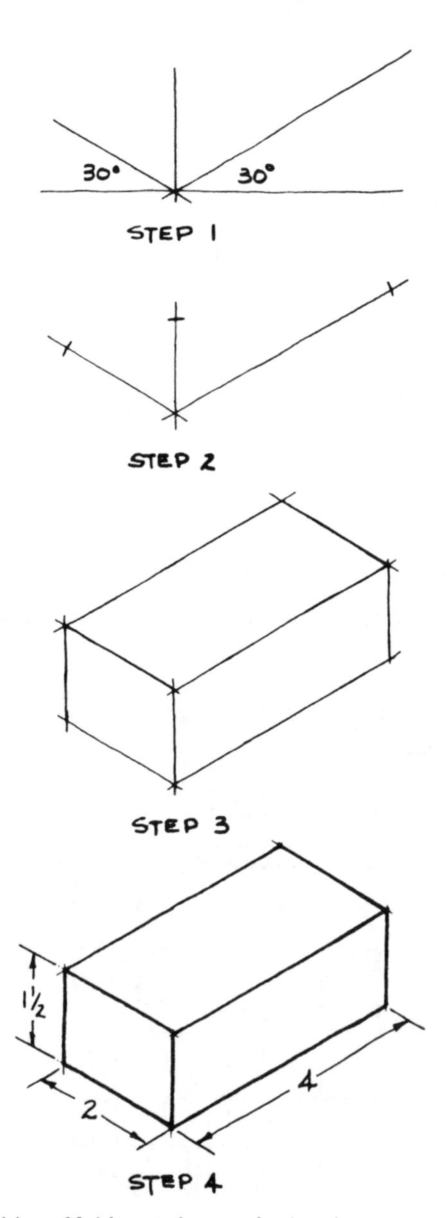

Figure 8-11 Making an isometric sketch.

Isometric Sketching

The principles of pictorial and orthographic sketching are similar, except that pictorial sketches are three-dimensional instead of two-dimensional. An isometric sketch is a pictorial drawing whose basic purpose is to show all three dimensions of an object in one sketch, instead of in multi-view orthographic sketches. Whereas orthographic sketches show the true size and shape of an object, pictorial sketches show the object as you would view it.

To make an isometric sketch, you need a 30-60 deg triangle, unless you are using isometric paper. First select the one view that shows the features of the object best. Either move around the object, or move it around until you can see everything you want to show. If the object is an idea in your mind, or if you are going to sketch an isometric view from an orthographic drawing, you will have to mentally visualize the object and assume a viewing position. Remember to start by sketching the three isometric axes 120 deg apart, using two angles of 30 deg, and a vertical axis of 90 deg. For example, to make an isometric sketch of a rectangular block measuring 1½ in. x 2 in. x 4 in., follow the step-by-step procedure illustrated in Fig. 8-11:

(1) Sketch the three isometric axes as described in the paragraph above and as shown in Step 1.

(2) Mark off the 1½ in. height on the vertical axis, the 2 in. width along the left axis, and the 4 in. length along the right axis, as shown in Step 2.

(3) Draw two vertical lines 1½ in. high (starting with the marks on the right and left axes) as shown in Step 3. Then sketch parallel lines from each of the marks on the sketch. (Note that the lines that are parallel on the object are also parallel on the sketch.)

(4) Place the dimensions parallel to the ends or edges as shown in Step 4.

Scale

Except to meet special requirements, sketches are not usually made to any scale. However, try to sketch objects in their true proportions as accurately as your eye can estimate. Do not try to mark off dimensions with a scale or rule on a freehand sketch. The size that you make the sketch depends only on the shape of the object and the requirements of the sketch.

SUMMARY

Good sketching skills are an asset to a journeyman. The ability to sketch quickly and accurately is useful when discussing your work with others, devising possible changes in a job, or keeping a record of the setup on a successfully completed job that may periodically recur.

Follow the tips on sketching techniques in this chapter and the rules of orthographic projection to produce good working sketches that can be used over and over. Remember that you are communicating information. Try to make your sketches clear and concise.

TRAINING PRACTICE

Given the following drawing, make a complete working sketch. Include all necessary views, dimensions, and notes.

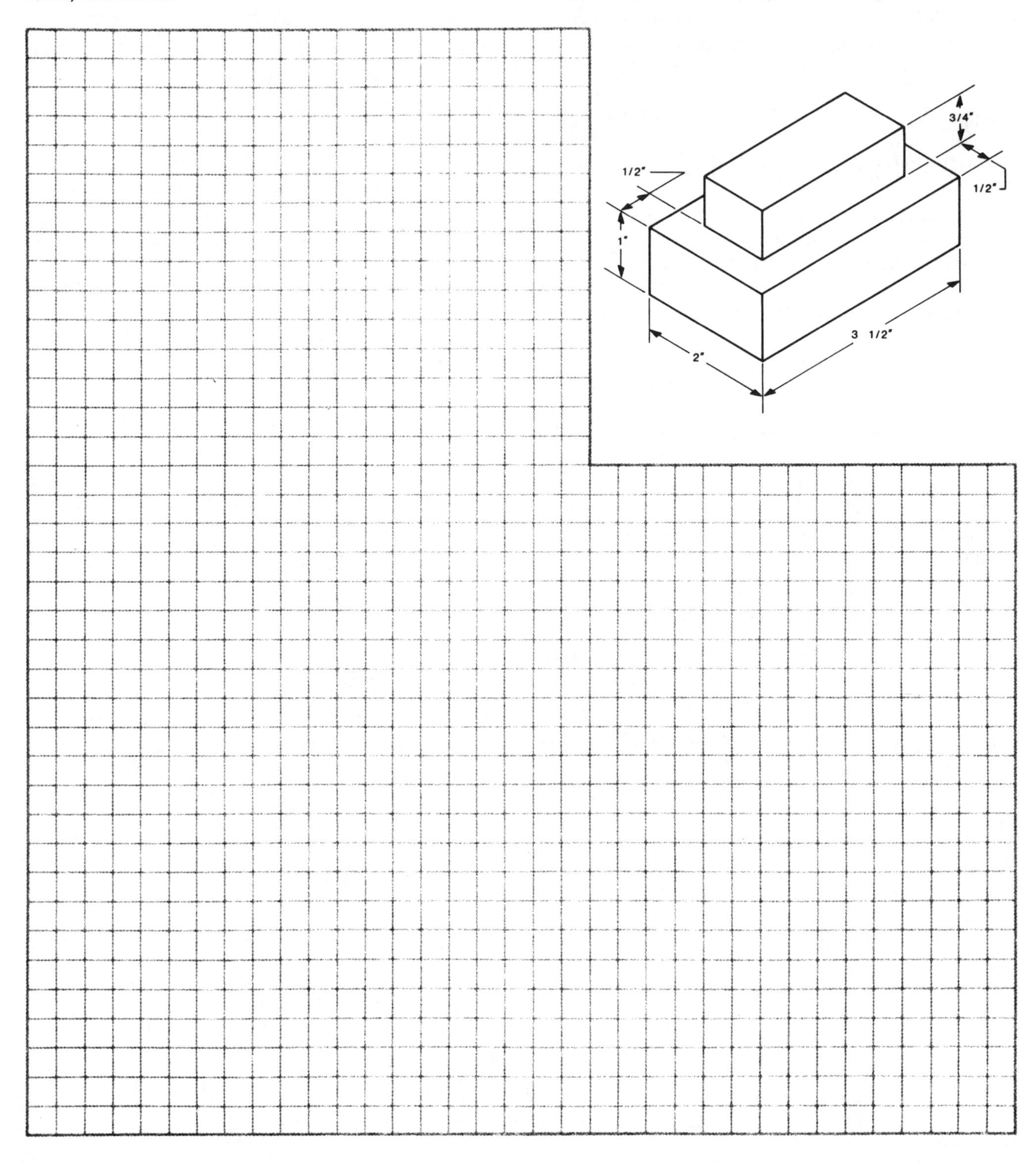

Chapter IX

LAYOUT

Many times in your job, you will be required to lay out work prior to machining it. Laying out the work means to plan the work on the surface of the material. It consists of scribing (marking) the lines that indicate the boundaries, centers, and other features on the object, so that you can machine it to the desired size and shape. The care that you take when you do the layout will determine the accuracy of the finished work.

WORK SURFACES

To perform accurate layout, measurement, and inspection in the shop, you must work on some type of smooth and level surface, such as a milling machine table or a surface plate. Surface plates are usually made from cast iron or granite that has been carefully machined or hand-scraped on the top. They are sometimes equipped with threaded inserts used to secure the workpiece and/or surface plate accessories. Several accessories are used with surface plates, such as parallels, V-blocks, and angle plates.

Surface plates are very expensive. Therefore, take care to protect them from damage. Make sure that the parts to be laid out are clean and free from burrs. You can remove high spots by grinding or filing. Make sure that any spatter or slag on flame-cut parts is chipped off. In addition:

(1) Do not allow anything to drop on the surface plate.

(2) Place only those items that you need to do layout on the plate. Never lay wrenches, hammers, etc., on its surface.

(3) Do not perform any hammering or punching operations while the work is on the plate.

(4) When the plate is not in use, apply a light film of oil to its surface to prevent rust. Cover the surface plate with a heavy felt pad followed by a sheet metal or wood cover to prevent accidental damage to the plate.

LAYOUT COMPOUNDS

In laying out work, you actually scribe the lines on a layout compound that has been applied to the surface of the work. Layout compounds provide a glare-resistant film on the work, and lines scribed through the film show up clearly. There are various colors and compositions of layout fluid. A commercial layout blue dye is the most common, but white and red are available as well. Dykem red and blue are used most often on surfaces that have been machined or have shiny finishes. White ink or lime and water are used on dark surfaces, like castings and weldments. Common chalk is sometimes used to lay out rough finished surfaces.

Because layout fluid evaporates rapidly, keep the container tightly closed when you are not using it. Apply a thin coating to the work, because compounds tend to flake off or produce thick or ragged lines when they are applied too heavily. Whichever type of compound you use, keep the surface clean and free from oil. Remove all burrs with a file or oilstone to prevent inaccurate measurements.

LAYOUT TOOLS AND USES

No matter what type of layout you do, or what the final object is to look like, you will use layout tools. The shape of the object and the accuracy required determine which tool to select. The following tools are the ones most often used.

Parallels

Parallels are bars of steel or cast iron that are machined so that their opposite surfaces are parallel to each other. Adjacent surfaces are usually at right angles to each other, but parallels are available with adjacent surfaces at other angles. Parallels are used when projections on the work prevent setting it directly on the surface plate, or when you want to raise items above the surface and still maintain parallelism.

V-Blocks

V-Blocks are used to hold round stock, and to prevent it from rolling or moving around. They have a 90 deg V-shaped slot cut into the top and bottom, usually of different depths as shown in Fig. 9-1. You place the stock in the slot lengthwise, and secure it with a clamp, if necessary.

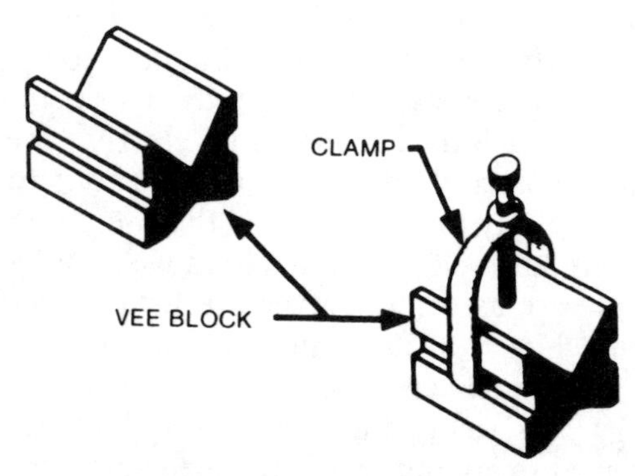

Figure 9-1 V-Blocks.

Angle Plates

Angle plates have accurately-machined surfaces at right angles to each other. When you want to mount the work at right angles to the surface plate, you can clamp it against the angle plate as shown in Fig. 9-2.

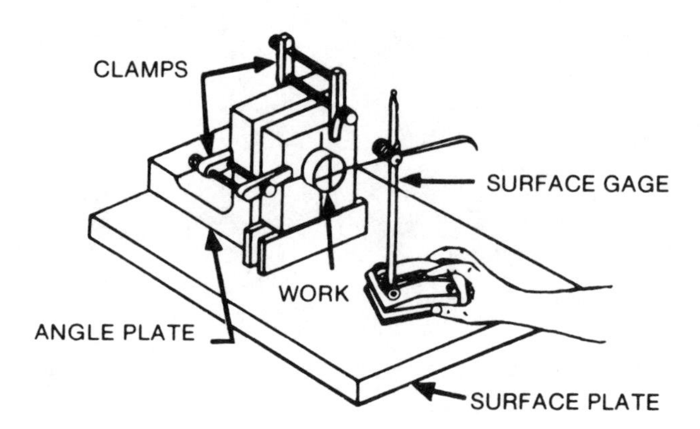

Figure 9-2 Using an angle plate in layout.

Scribers

Scribers are tools with sharp points on both ends (see Fig. 9-3) used to mark layout lines through the film of layout compound. You should keep the points of the scriber sharp with an oilstone, so that you can draw thin, accurate lines. Make only one line with the point of the scriber, as shown in Fig. 9-4. Do not go back over the line with the scriber, because this usually results in a series of closely-spaced lines that are not accurate. Also, don't exert too much pressure on the scriber, or you will scratch the surface of the work.

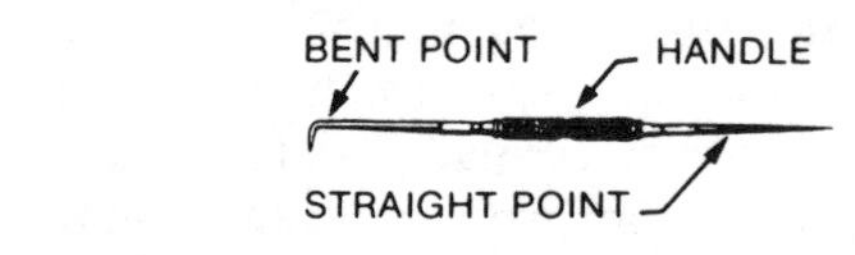

Figure 9-3 Scriber.

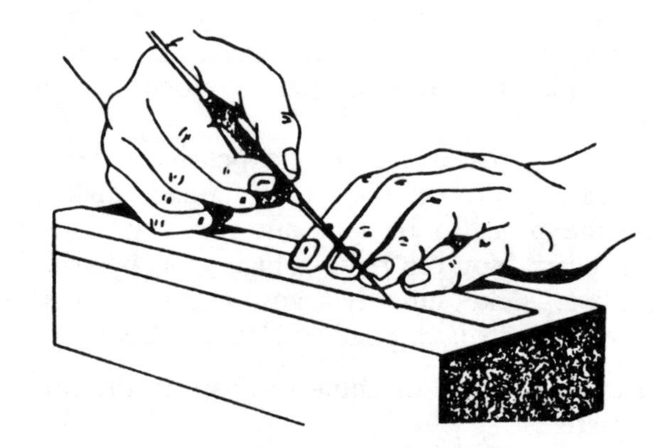

Figure 9-4 Using a scriber to draw a line.

Combination Set

The combination set, shown in Fig. 9-5, consists of four units: square head, protractor head, center head, and steel rule, also known as the *blade.* The blade is the basic unit. It has straight, smooth edges to help you draw straight lines with the scriber, and a scale that is accurate to 1/64 in. The other three units attach to the blade. They slide along it, and can be clamped rigidly in any desired location.

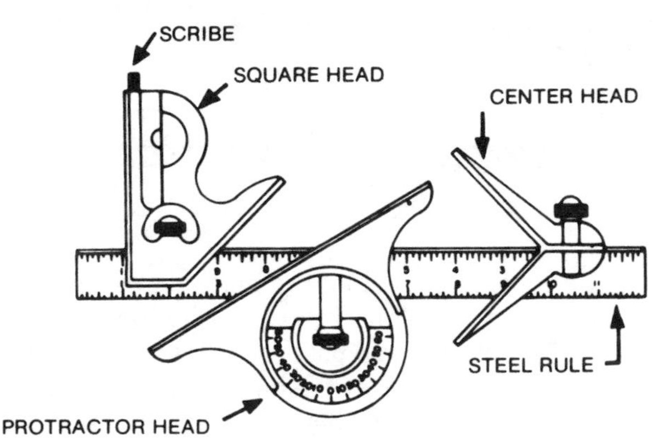

Figure 9-5 Combination set.

By clamping the **square head** to the blade, you can scribe lines parallel to an outside edge, and scribe perpendiculars to those lines or edges, as shown in Fig. 9-6A. The other edge of the square head forms a 45 deg angle with the blade for use when you need to scribe or check such an angle, as shown in Fig. 9-6B.

You will use the **protractor head** to measure or lay out angles other than 45 deg or 90 deg, as shown in Fig. 9-7.

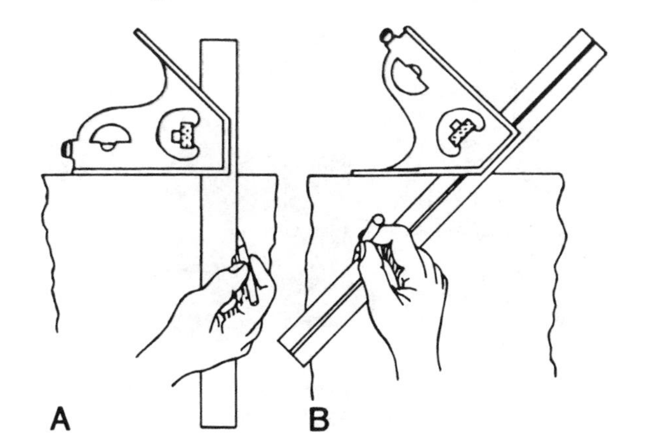

Figure 9-6 Using the square head.

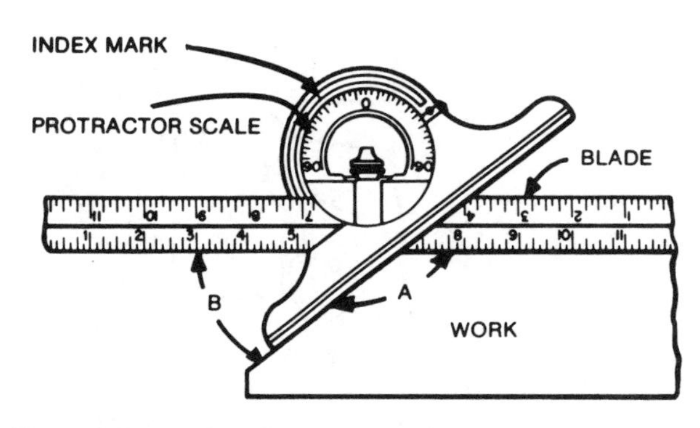

Figure 9-7 Using the protractor head.

The **center head** provides a fast way to locate the center of cylindrical objects by first clamping the center head to the blade, and then placing the V-shaped opening snugly against the work as shown in Fig. 9-8. When you scribe a line along the inner edge of the blade, it will pass through the center of the work. By rotating the center head or work 90 deg and scribing a second line in the same manner, you form an intersection that marks the exact center of the cylindrical piece.

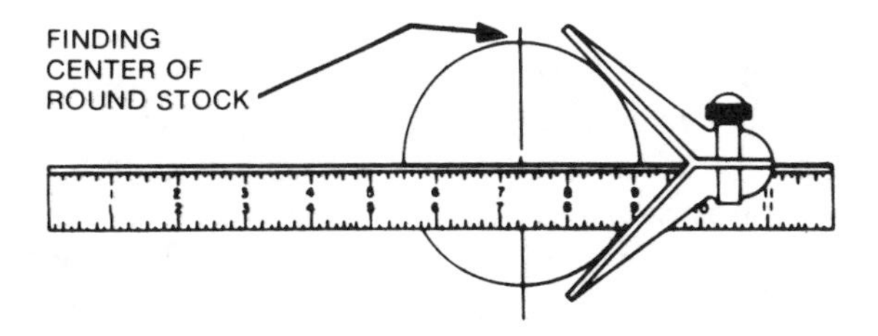

Figure 9-8 Using the center head.

Machinist's Square

The machinist's square (sometimes called *machinist's try square*) is used to lay out or check perpendiculars when you require more accuracy than the combination square head and blade provide. Figure 9-9 shows how to use the square to test adjacent faces of a piece of work for squareness.

In manufacturing this tool, the edges of both the body and the blade are carefully machined and then lapped to exact trueness. This precision tool will remain accurate only if you take special care of it. Do not allow it to contact other tools that could dent or nick its lapped edges or distort its 90 deg setting. Clean and polish it frequently to keep it from rusting, and store it in a safe place when you are not using it.

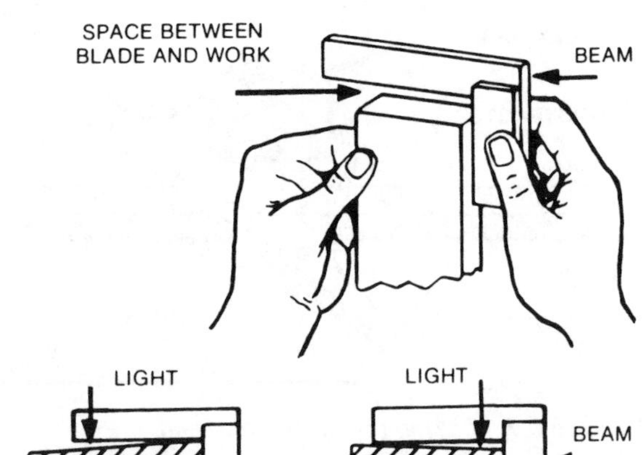

Figure 9-9 Using the machinist's square.

Surface Gage

The surface gage is an adjustable scriber that allows you to scribe horizontal layout lines as shown in Fig. 9-2, to locate centers in rough work, and can also be used as a height gage for leveling work on a machinist's vise or plate. Because you can adjust and lock both the vertical spindle and the scriber in any angular position, you can scribe layout lines on a piece at any given height and from almost any position.

To use the surface gage to scribe horizontal lines, first set the scriber point to the desired height on the blade of the combination set, using the square head to hold it in the vertical position. For accurate settings, use the thumbscrew located near the rear of the surface gage base to obtain fine movement of the point. Then scribe the horizontal line, as shown in Fig. 9-10.

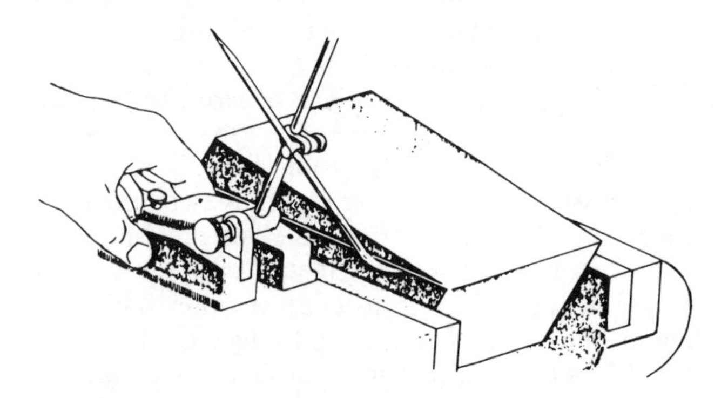

Figure 9-10 Using the surface gage to scribe horizontal lines.

The surface gage, like the machinist's square, is a precision instrument. It requires periodic maintenance. Oil all moving parts so that they operate freely. When you are not using it, store the surface gage carefully in a safe place to avoid damaging it.

Vernier Height Gage

When a high degree of accuracy is necessary, you will use the vernier height gage (see Fig. 9-11) instead of the surface gage. The vernier height gage allows you to set the scriber point accurately to 0.001 in. You read it the same way as the vernier caliper.

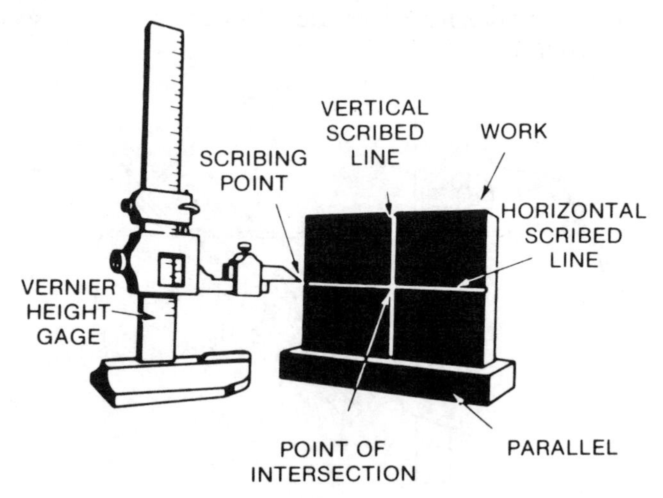

Figure 9-11 Using the vernier height gage.

Layout Punches

Although your toolbox may contain many different kinds of punches, only the prick punch and the center punch are used as layout tools. The prick punch (see Fig. 9-12) is ground to an included angle of 30 deg. The long, slender point allows you to position it accurately on the layout lines. Use it to make fine punchmarks along scribed lines so that you can find them even if the lines wear off, to locate centers of holes, and to provide a pivot point for the leg of the divider (see below) when laying out holes or radii.

The center punch, also shown in Fig. 9-12, differs from the prick punch in that it has an included angle of 60 deg. Use it mainly to make prick marks deeper prior to drilling. Keep the points of these punches sharp by grinding them frequently. Check the driving end of the punch at the same time to make sure that it is flat and perpendicular to the centerline of the punch. Remove any dangerous burrs that form around the edges.

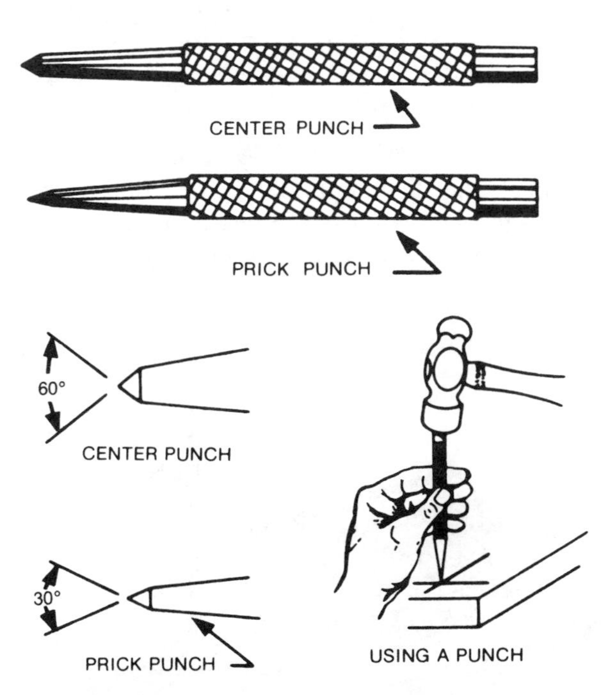

Figure 9-12 Center punch and prick punch.

Dividers

You will use spring-jointed dividers (see Fig. 9-13) to transfer distances, and to scribe arcs and circles on the work surfaces. Inspect the divider points often to make sure that they are sharp enough to make a clean, precise scratch on a metal surface. If the points are dull, sharpen them on an oilstone by rotating the tool between your forefinger and thumb as you rub the points back and forth against the stone.

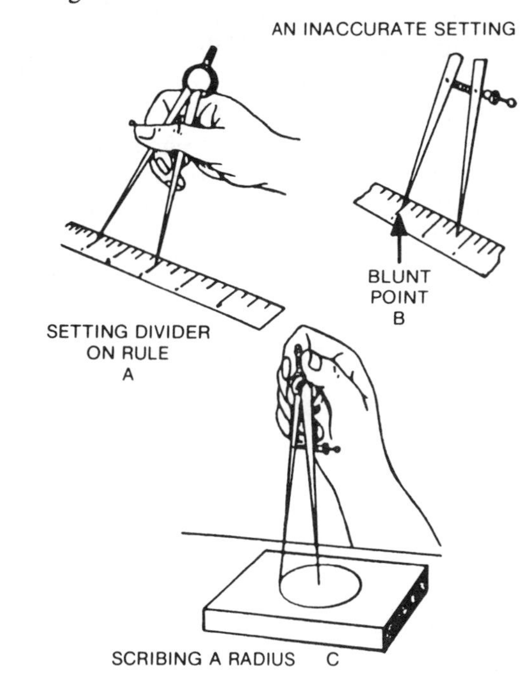

Figure 9-13 Using dividers.

Trammel Points

Use trammel points, shown in Fig. 9-14, when you must mark a distance or radius larger than you can with dividers. You first position the points anywhere along the rod to the proper distance, and then use this tool as you would dividers.

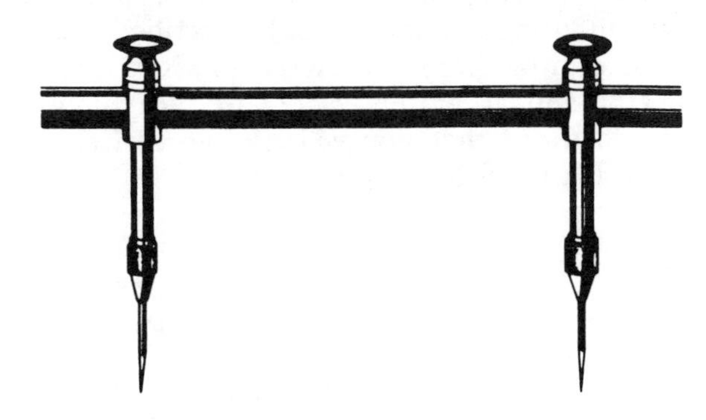

Figure 9-14 Trammel points.

Hermaphrodite Calipers

Although most calipers are considered measuring instruments, the hermaphrodite caliper is also used as a layout tool. It is so named because it has one straight sharp-pointed divider leg and one curved (or bent) dull-pointed caliper leg. You can adjust these legs to either outside or inside measurements, and scribe lines parallel to edges as shown in Fig. 9-15. You can also use this caliper to locate the approximate center of a boss or other round projection on a piece of work. To do this, first set the caliper to roughly one-half the diameter of the boss, and then scribe four arcs spaced approximately 90 deg apart. This forms a small square around the true center of the boss. Do not use this method when great accuracy is required, however.

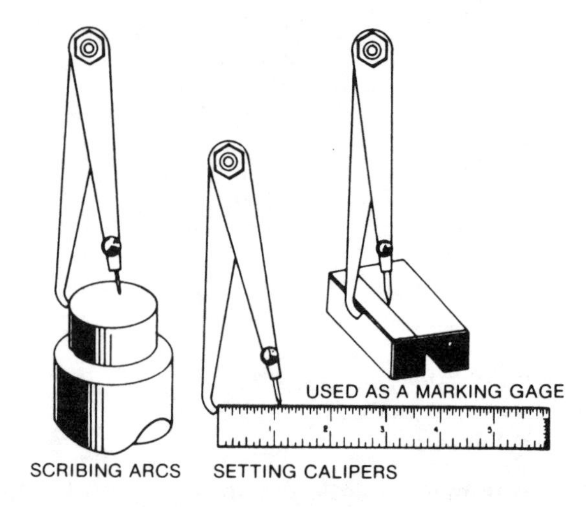

Figure 9-15 Using hermophrodite calipers.

LAYOUT METHODS

Before cutting material for a job, you will lay it out using the layout tools described above. No matter how simple the layout job is, try to do it neatly and accurately. Some of the common layout methods that you will use most often are described in the following sections.

Holes and Radii

The center of holes and radii are located by intersecting lines–vertical and horizontal lines, horizontal and angular lines, etc.–or by intersecting arcs. In short, the point where any two lines cross can be a center.

(1) Use a prick punch to mark the center at an intersection. This will provide a spot on which to pivot the leg of the divider without "walking" off of location.

(2) Set the divider for the proper radius, and scribe the arc or circle desired.

(3) Use the prick punch to make a series of small, equally-spaced indentations on the circular scribe mark to show the outline more clearly.

Shaft Features 90 Deg Apart

You can lay out keyways and holes positioned 90 deg apart on a shaft simply and accurately using the following procedure illustrated in Fig. 9-16. For two keyways at 90 deg to each other:

(1) Place the shaft on a V-block for easy turning.

(2) Adjust the surface gage to the center of the shaft.

(3) Scribe a line across the center of the end of the shaft, as was shown in Fig. 9-2. Also scribe a line along the side of the shaft in the position of the part feature.

(4) Rotate the shaft 90 deg, using a square to check the alignment, and scribe a line (with the surface gage) across the end and along the side of the shaft for the second keyway.

Figure 9-16 Laying out keyways and holes on a shaft.

Hexagons

Six-hole bolt circles are common in layout. They are really only hexagons. The points of the "hex" serve as the hole centers. Referring to Fig. 9-17:

(1) Scribe a circle with the proper bolt circle diameter.
(2) Punch a starting point at some convenient location on the circle.
(3) Using dividers set to the bolt circle radius, place one leg of the divider on the starting point and mark off two other points. Punch these points with a prick punch.
(4) Mark two new points, and punch as above. Keeping the dividers set at the proper radius, continue marking additional points, using the previously-marked and punched points as references. The last point is a check on yourself. The scribe marks from last two points should cross exactly at the same place on the circle.

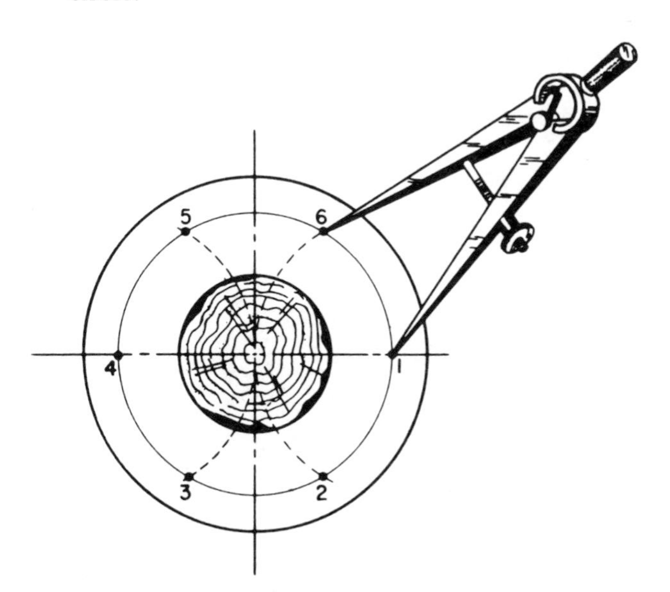

Figure 9-17 Laying out a hexagon.

Octagons

Another common bolt pattern is eight-hole, or an octagon. An octagon is simply two squares laid on top of one another at 45 deg. Referring to Fig. 9-18:

(1) Construct the correct size bolt circle.
(2) Draw a centerline through the circle.
(3) Bisect the diameter, giving four points.
(4) Bisect the distance between adjacent points. This gives a total of eight hole centers.

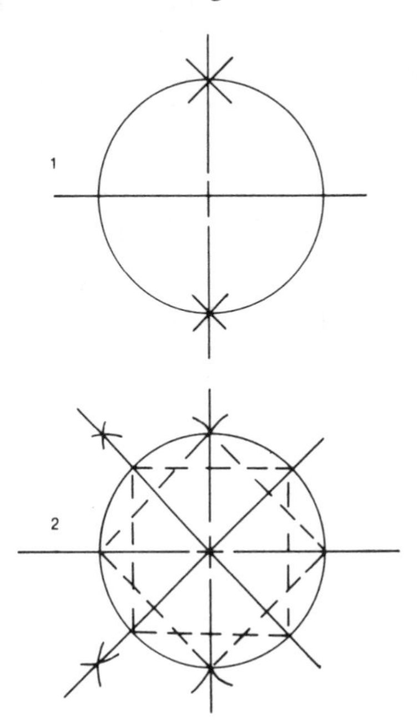

Figure 9-18 Laying out an octagon.

Finding the Center of Round Stock

You can use the hermaphrodite caliper to find the center of a radius or the center of a circle as shown in Fig. 9-19.

(1) Set the hermaphrodite for approximately one-half the diameter.
(2) Keeping the same setting, hook the curved leg of the caliper on the edge of the work, and scribe four arcs, 90 deg to each other.
(3) Where the four arcs converge is the center. Punch them as described above.

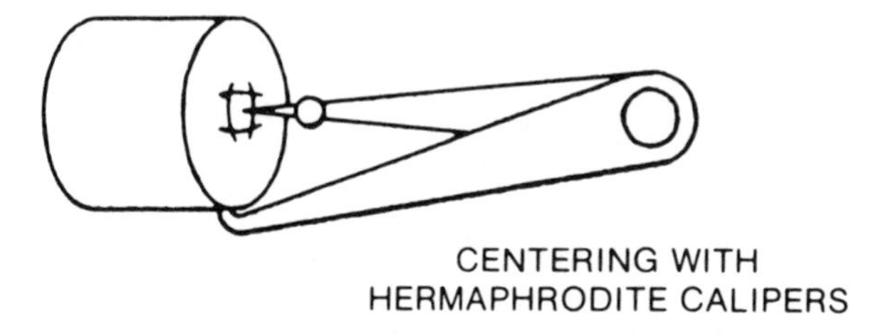

Figure 9-19 Finding the center of round stock.

Bridging a Bore

Any job with large diameter holes already bored in it (castings, weldments, partially machined work, etc.) may require that you take some dimension from the center of a bore. "Bridging the bore" is the name of a technique that will give a center where one is needed, but when there is nothing but a hole to put one on.

(1) First, fit a fine-grained wood plug into the hole, as shown in Fig. 9-17, or cut a 1 in. wide strip of wood to fit the diameter of the bore.

(2) Use the hermaphrodite caliper as instructed above to find the approximate center.

(3) After locating the center, use dividers to scribe a true circle, and check the location.

SUMMARY

A layout means to scribe the boundaries, centers, and feature locations on a workpiece before you machine it. Final positioning of features is accomplished using accurate measuring tools. The layout itself is a reproduction of the print on the actual part surface. You must know how to read a blueprint correctly to transfer the information to rough stock.

This chapter describes the layout tools and simple techniques that will help you perform many kinds of layout operations.

TRAINING PRACTICE

Use a straight edge/scale and compass (not dividers) to perform the following exercises.

1. Construct a 3 in. bolt circle with a 6-hole pattern.

2. Construct a square with 2 in. sides.

3. Construct a 3-hole pattern on a 2 in. bolt circle.

4. List, in sequence, the steps needed to construct the centerlines of two keys on a shaft. The keyways are at opposite ends of the shaft and 90 deg apart.

Chapter X

BASIC BLUEPRINT READING

Chapters X and XI will give you the opportunity to use what you have learned from the earlier chapters of the book, and to read actual shop blueprints. These prints were taken directly from industry, and represent what is typical in the machine trades. On the job, you will be asked to read similar prints to create the parts that your employer wants to produce.

Each set of questions applies to a particular drawing or drawings. Space is provided for your answers. In many cases, a short description of the part precedes each set of questions to help you orient yourself to the print. To begin, study the print for a few moments to acquaint yourself with the location of views and technical information. Try to determine during that time what the part is, what it looks like in three-dimensional space (use your visualization skills), and try to deduce what its function might be.

After this brief orientation, use any mechanical knowledge you have to establish a machining process for the part. Try to determine which operations should be performed first, second, third, etc. If you get into the habit of approaching blueprint reading in a systematic manner, you can avoid many careless mistakes in reading blueprints and machining parts.

TP-100

Description: The part is a typical 12-32 nut for a machine screw.

Answer the following questions:

1. What is the shape of the nut? __________
2. Name the views. __________
3. What does ".375" mean? __________
4. What material is the nut made from? __________
5. How will the 12-32NS2 threads be made? __________
6. What type of heat treating will be used on the nut? __________
7. What is the dimension of the relief diameter? __________
8. Give the maximum and minimum acceptable diameters of the relief. __________
9. What does the hidden line represent? __________
10. What is the overall thickness of the nut? __________
11. Have any revisions been made to this print? __________
12. What is meant by the note "*Stock*"? __________
13. What is the maximum angle of the chamfer? __________
14. What is the fractional and decimal pitch of the thread? __________
15. Where does the ± .003 tolerance apply? __________

TP-101

Description: TP-101 is a three-view drawing of a small shaft. The part is characteristic of a lathe job, with some necessary milling. The flat on the left end indicates that the part may be driven from that end. The right end, because of its close tolerance on the small diameter and its length, indicates that it may be pressed into a bearing or other mechanical member.

Answer the following questions:

1. What kind of material is the shaft made from? ______________________________

2. How large is the actual part compared to the drawing? ______________________________
____

3. Name the views in the drawing. ______________________________
____

4. What is the overall length of the part? ______________________________

5. What type of coating will the part receive? ______________________________

6. What are the maximum and minimum acceptable sizes of the .156 dia? ______________________________
____

7. What are the maximum and minimum acceptable sizes of the .124 dia? ______________________________
____

8. How deep is the cut needed to produce the flat? ______________________________

9. What is the distance from the shoulder on the flat to the shoulder on the .124 dia? ______________________________
____

10. What are the accepted maximum and minimum sizes and angles of the chamfers? ______________________________
____

11. What is the length of the flat? ______________________________

12. What is meant by ".156 stock"? ______________________________

TP-101 (cont.)

13. Name Line Ⓐ. ______________________________

14. What type of tolerancing is Ⓑ? ______________________________

15. What is the mean dia of Ⓑ? ______________________________

16. What is the maximum overall length of the shaft? ______________________________

TP-102

Description: The part is a straight aluminum tube with chamfers and inside threads on both ends.

Answer the following questions:

1. What shape and type of material is specified? ______________________________

____

2. What are the ID and OD dimensions of the tubing?
   ID = ______________________ OD = ______________________

3. What tolerances are stated for three-place decimals? ______________________

4. Given 1.275-32NS-2LH, what do the following mean?
   a. 1.275 ______________________
   b. 32 ______________________
   c. NS ______________________
   d. 2 ______________________
   e. LH ______________________

5. How deep are the threads? ______________________

6. What are the maximum and minimum sizes of the finished length? ______________________

____

7. What do lines (A) represent? ______________________

____

8. What do the lines at (B) represent? ______________________

____

9. What type of view is given? ______________________

10. Sketch a right side view with all necessary lines (omit dimensions).

TP-102 (cont.)

11. What is the scale of the drawing?____________________

12. Would you turn the part clockwise or counter-clockwise to tighten it?____________________

13. Which dimensions does the 3 place ± .005 apply to?____________________

    ____

TP-103

Description: This blueprint gives three views of a shaft. The shaft is made for rotary motion instead of reciprocating motion, due to the flange and bolt holes at one end, the keyway at the opposite end, and the $\frac{1.0007}{1.0003}$ diameters apparently turned for bearings or some part fitted over the shaft and locked into position by the .252 holes.

Answer the following questions:

1. Will all sharp edges need to be broken? ______

2. What surface finish is required in general? ______

3. What is the mean dimension of the .375 $^{+.000}_{-.001}$? ______

4. What dia is the C'drilled hole? ______

5. What is the smallest radius allowed in the .090R max dimension? ______

6. What material is the part made from? ______

7. How hard will the part be after heat treatment? ______

8. What can be said about the $\frac{1.0007}{1.0003}$ and the .8748 dia in the section view relative to one another? ______

9. What is the depth of the keyway? ______

10. What kind of dimension is $\frac{.249}{.251}$ in the right view? ______

11. Describe a 30√ (microfinish). ______

12. Given the note "1/2-20 TAP-1.000 DP" in the section view, explain the following:
    a. 1/2 ______
    b. 20 ______
    c. 1.000 DP ______

TP-103 (cont.)

13. What would you call the .937 dia in the section view? ____________________

____

14. What is the depth of the .453 dia hole in the section view? ____________________

____

15. What is the depth of the .500 dia hole in the section view? ____________________

____

16. Approximately how long is the .8748 dia? ____________________

17. How deep are the .252 dia holes in the section view? ____________________

____

18. How many chamfers will this part receive? ____________________
How many C'sinks? ____________________

19. What is the maximum acceptable overall length of the part? ____________________

____

20. What is the diameter of the bolt circle? ____________________

21. Looking at the right view, list below the diameters around the centerline starting with the smallest:

Smallest 1. .453 *(Example)* ____________________

2. ____________________

3. ____________________

4. ____________________

5. ____________________

6. ____________________

7. ____________________

Largest 8. ____________________

TP-104

Answer the following questions:

1. What is the name of the part? ____________________

2. What kind of coating will the part receive? ____________________

 ____

3. How many holes are on the 2.270 basic? ____________________

4. How close to the given position are the holes in Question 3 to be located? ____________________

5. What microfinish is required in the .733 to .737 Ø holes? ____________________

 ____

6. What type of material is the part made from? ____________________

7. What is the overall thickness of the part? ____________________

8. What is the mean dimension of the O.D. of this part? ____________________

9. What is the tolerance on the 2.270 dia? ____________________

10. What is the name of this sectioning method? ____________________

11. What is the meaning of the centerlines of the holes on the bolt circle? ____________________

12. What microfinish is required on the thickness of the part? ____________________

13. What is the general microfinish required on surfaces not specified? ____________________

 ____

14. Assume that the I.D. of the part is 1.373 before plating. The plate shop adds .001 of plate. What will the dia of the holes be after plating? ____________________

 ____

TP-105

Answer the questions below:

1. What size piece of stock would you cut for this part?
   A) Dia ______________ B) Length ______________

2. What is the depth of the slots? ______________

3. How do the threads at either end of the bore differ? ______________

   ____

4. Which is more critical, the horizontal center distance of the 3/32 holes in the front view, or their location from the end of the part? ______________

   ____

5. Which is more critical, the width of the 3/4 slots or the location of the shoulder created by the slots?

   ____

   ____

6. Which kind of dimensions would the 1/64 tolerance apply to? ______________

   ____

7. What size radius is used to scallop the ends of the part at (1)? ______________

   ____

8. What kind of material is this part made from? ______________

9. Given the notation: .75-16NF-3B × 1.625DP, identify the following:
   a. 75 ______________
   b. NF ______________
   c. 3 ______________
   d. B ______________
   e. 1.625 ______________

10. What size tap drill should be used for the .75-16NF thread? ______________

    ____

11. How deep should the 1/16 holes be drilled? ______________

TP-105 (cont.)

12. What feature do the hidden lines at (2) represent? ______________________

____

13. What feature do the hidden lines at (3) represent? ______________________

____

14. What is the dimension across the flats of the 3/4" slots? ______________________

____

15. Are the 3/4" slots centrally located in the front view? ______________________

____

16. How many 1/4 radius pockets are there? ______________________

____

TP-106

Description: The object in TP-106 is a weldment called a discharge pipe. A weldment is any object fabricated by welding the separate parts together to form one complete unit. In this case, the resulting object is a pipe fitting consisting of a 90° elbow, a flange, and a length of threaded pipe. This assembly will be attached to a major part by seven screws. An "O" ring will be used to seal to prevent leakage.

Answer the following questions:

1. Name both views. ______________________

2. How many pieces are included in the weldment? ______________________

____

3. Are there threads on the O.D. of the pipe? ______________________

4. What is the size of the bolt circle? ______________________

5. What does N.P.T. mean? ______________________

____

6. Will the 5 in. length of pipe be welded to the elbow first, or will the 90° elbow be welded to the flange first? ______________________

   How do you know? ______________________

   ____

7. In the top view, why are phantom lines shown on elbow? ______________________

____

____

8. How many holes are on the bolt circle? ______________________

9. What dimensions and angles are used to locate the holes? ______________________

____

10. How deep is the groove? ______________________
    How wide is the groove? ______________________

TP-106 (cont.)

11. Explain why the face of the flange and the groove need a smooth finish. ____________________

    ____

12. According to the print, when should the flange be machined? ____________________

    ____

13. What is the overall length of the assembly? ____________________

    ____

TP-107

Description: This crank is a very intricate part. The width, thickness, and overall center-to-center length are critical dimensions.

Answer the following questions:

1. What is the overall length of the object? ____________________

2. What is the overall width of the part? ____________________

3. What kind of material is the part made from? ____________________

4. How many #30 drilled holes are there? ____________________

5. What size countersink is required? ____________________

6. How far from the .7488-.7493 dia is the center for the .437 radius? ____________________
   Vertical? ____________________ Horizontal? ____________________

7. What is the significance of the .437 dimension and the phantom line at (A)? ____________________
   ____

8. What does the term "line ream" indicate? ____________________
   ____

9. Will this part be checked for cracks? ____________________
   ____

10. What is the length of the slot? ____________________

11. Are there any surfaces that receive a smoother finish than 125√ ? ____________________
    ____

12. What is the radius of the part at the small end? ____________________

13. How far off-center from the .7488-.7493 dia is the #30 drilled hole? ____________________
    ____
    ____

14. Which would you machine first, the outside surfaces of the legs or the inside groove?
    ____

TP-108

Description: Because the part name is "adaptor," we can assume that its function is to facilitate the mounting of two dissimilar parts on a common assembly.

Answer the following questions:

1. Give the overall dimension of the part:
   Thickness: ____________ Width: ____________ Length: ____________

2. What kind of material is the part made from? ____________

3. What does the abbreviation H.R.S. mean? ____________

4. Name the three views on the drawing.

   a. ____________

   b. ____________

   c. ____________

5. Where does the cutting plane lie? ____________

   ____

6. The $2^1/_2$ dia in the right section view represents which kind of bore? ____________

   ____

7. What does the phrase "far side" mean in the #8-32 tap note? ____________

   ____

8. What size is the radius at "R. TYP. 4 PLCS"? ____________

9. Given that the slots are made to adjust the adaptor side-to-side, how much total movement is possible if the diameter of the inserted stock is 1.030? ____________

   ____

10. What is the size of the pipe tap used on this drawing? ____________

    ____

11. Which machining operation will be used to form the L-shaped #0 drilled hole? ____________

    ____

TP-108 (cont.)

12. What are the dimensions of the rectangular pad projecting from the back side of the adaptor?
    Depth of Protrusion: ____________________ Width: ____________________
    Length: ____________________

13. What is the angle between the #8-32 tapped holes? ____________________

14. What dimension is used to inspect accuracy of depth in 2½ Ø C'bore? ____________________
    ____

15. What is the depth of the 2½ dia bore measured from surface ①? ____________________
    ____

16. Give the reason for the small view on the left. ____________________
    ____

17. Calculate the length of the elongated slot using basic dimensions. ____________________
    ____

18. What are the locating dimensions for the four .343 Ø holes?
    Horizontal: ____________________ Vertical: ____________________

19. What are the center distances between the four .343 holes?
    Horizontal: ____________________ Vertical: ____________________

20. Will it be necessary to machine surface ①? ____________________
    ____

TP-109

Description: The block in TP-109 is a precision die casting, which is produced in a die cast die. Such dies are made from steel, and injected with molten metals like zinc, aluminum, etc. The draft angle is necessary to aid in the easy ejection of part from die. 1/32 must be left on surfaces to be finish machined.

Answer the following questions:

1. What kind of material is the part made from? ____________

2. Give the proper name of the dimensioning plan used on this print. ____________

3. Explain the reason for the zigzag in the 16.910R leader. ____________

4. How deep is the large pocket recessed below the part surface? ____________

5. How many .641 dia holes are in this part? ____________

6. What is the overall thickness of the part? ____________

7. What is the size of the radius running parallel to the 16.910R? ____________

8. How did Revision A affect the depth of the pocket in Section A-A? ____________

9. Will any of the surfaces of this part be finished machined? ____________

10. How deep will the .641 dia holes be made? ____________

11. How far from the horizontal datum line is the widest portion of the part? ____________

12. What kind of surface treatment will this part receive? ____________

TP-109 (cont.)

13. What is the thickness of the rim on the right? ______________________________

14. What is the X dimension? ______________________________

15. What is the Y dimension? ______________________________

16. What is the vertical center distance between each of the .641 holes? ______________________________

    ____

17. If the draft angle were shown in Section A-A, which dimension would be larger: the mouth of the pocket or the base of the pocket? ______________________________

    ____

18. Starting with hole ① and going clockwise, list the vertical and horizontal locating dimensions for each of the .370 Ø holes.

	Horizontal	Vertical
1)		
2)		
3)		
4)		
5)		
6)		
7)		
8)		
9)		
10)		
11)		
12)		
13)		
14)		
15)		

TP-110

Description: To help visualize this casting and its use, the parts drawing (upper left) is given. If you try to make sense of how things work and why, your ability to interpret drawings and machine parts will improve.

Answer the following questions:

1. How many bores require a 16√, and what are their diameters? ______________________

2. Give two ways that these bores are critical. ______________________

3. What is the maximum that the $\frac{2.0472}{2.0474}$ dia can be out-of-parallel with Surface (B)? ______________________

4. Explain the purpose of the two .3125 dia rolls. ______________________

5. What is the maximum allowable distance between rolls? ______________________

6. Why is the hole in Section AA elliptical? ______________________

7. How many #6-32 tapped holes does this part contain? ______________________

8. What size tapped holes other than #6-32 will be used? ______________________

9. What is the diameter of the bolt circle at (1)? ______________________

10. In which sequence would you place the following major operations?
    - __________ Boring the $\frac{2.0472}{2.0474}$ dia.
    - __________ Machining Surface (B) and the dovetail
    - __________ Drilling and boring the series of diameters along Section A-A.

11. After the spot face has been cut, how deep should the $\frac{.562}{.560}$ bore be made? ______________________

TP-110 (cont.)

12. How far from Surface "B" is the bottom of the dovetail? ______

13. Assume that the dovetail will be cut on a milling machine with a dovetail cutter. At what angle will the cutter be ground? ______

14. With reference to Surfaces (2) and (3), is the dovetail slot on-center? ______

______

15. What size drill should be used first on the bores in Section A-A? ______

______

16. What size tap drill is required for a #6-32 tap? (*Hint:* Use *Machinery's Handbook.*) ______

17. Draw a 3 view sketch of the part that will fit in the 1/2 x 3/4 x 3/4 DP elongated slot.

18. What type of metal might be used to make this part? ______

19. Describe its use. ______

______

______

TP-111

Description: This pulley is designed for both flat and V-shaped belts.

Answer the following questions:

1. What is the part made from? ______

2. How much larger is the part than the size represented by the drawing? ______

3. What is the horizontal starting dimension on this drawing? ______

4. What type of finish mark is used? ______
   Describe the finish. ______

5. Do the spokes have a uniform cross section? ______

6. Is the entire drawing made to the same scale? ______

7. Why is the groove dimensioned 7/16 in one view, and .521 in the other? ______

8. What is the depth of the V-groove? ______

9. What is the width of the finish-machined V-groove at its widest point? ______

10. What are the maximum and minimum angles in the V-groove?
    Max. ______ Min. ______

11. What is the diameter of the counterbore? ______
    How deep is it? ______

12. What is the largest diameter on the part? ______

13. What is the finished diameter at the bottom of the V-groove? ______

14. How long is the 2.001 dia bore? ______

TP-111 (cont.)

15. What is the width of the spokes 3″ away from the centerline? ______________

16. How much do the spokes taper in $2\frac{1}{2}$″ of length?
    Width: ______________ Thickness: ______________

17. What is dimension "X"? ______________

18. What type of section view is the right view? ______________
    ____

19. In the partial view, is the .521 dim central? ______________
    ____

TP-112

Description: The bell crank is a part mass-produced on a progressive die. The material originally was .040 thick flat stock. During the manufacturing process, the holes were punched, the contours were cut, and the part was bent into a U-shape. All holes have a close tolerance, and their position is governed by the center top hole.

Answer the following questions:

1. What is meant by the note: "Case Hdn (sometimes abbreviated CH) .003-.006 deep Rc 60/65"?
2. How is the alignment of the .125 dia holes checked?
3. At what time during the manufacturing process should the part be case hardened?
4. What type of material is the part made from?
5. What is the dimension of the radius marked "R(3) places"?
6. What is the smallest hole in the part?
7. What is the thinnest metal that could be used to make this part?
8. What is the vertical center distance between the .094 ⌀ and the .083 ⌀ hole?
9. What is the largest radius shown?
10. Considering tolerances, what would the smallest possible dimension at ⊗ be?

TP-113, 113-1 & 113-2

Description: A pinch valve (not shown) will contact the large face of the plunger. A hydraulic cylinder (not shown) will be connected to the yoke. When the yoke moves slightly to the right (drop in pressure at cylinder) the plunger will move to the left, opening the pinch valve and allowing small amounts of fluid to balance the pressure. Fine adjustments to pressure can be made by rotating the eccentric pivot. The assembly will be held onto the machine with 1/4-20 screws.

Answer the following questions:

1. How many individual parts will make up this assembly? ____________________

2. How many parts does the sub-assembly consist of? ____________________

3. What is the purpose of the sub-assembly view? ____________________

4. What type of materials are the following details made from?
   Yoke __________ Bushing __________ Eccentric __________
   Plunger __________ Block __________ Post __________

5. If (9) was rotated one complete turn, how far would it travel? ____________________

6. How far will (19) protrude into (4)? ____________________

7. How far will (18) protrude from (17) after assembly? ____________________

8. What is the purpose of (20) ? ____________________

9. In (1), what is meant by S.F.? ____________________

10. In (1), what size will the .250 Ø be made to before heat treating? ____________________

11. In (2), why is the part hardened? ____________________

12. Why is "DR. POINT BREAK THRU PERMISSIBLE IN TWO HOLES" in the holder? ____________________

TP-113, 113-1 & 113-2 (cont.)

13. What is the purpose of the .3750 $^{+.0006}_{-.0000}$ phantom hole in Det. ④? ______________________________
____

14. In ④, what type tool will be used to produce the .187 P.F. hole? ______________________________
____

15. What detail number fits into the #4-40 threads in ⑩ ? ______________________________
____

16. In Det. ⑰ :

    A) What size and type cutter will be used to produce the elongated slots? ______________________________
    ____

    B) How long are the flats of the elongated slots? ______________________________

    C) What is the maximum depth of the .155 dia C'bore? ______________________________

    D) What is the purpose of the .031 Ø? ______________________________

    E) What does the abbreviation P.F. mean? ______________________________

    F) How will ⑯ and ⑮ be held into ⑰ ? ______________________________

17. If the .1405 and .3748 dias run true to the ℄ of Det. ⑫ , how much would the .2498 dia run out T.I.R.? ______________________________

18. Why does the face of ⑱ require a 32√ ? ______________________________

19. What Det. # fits in the I.D. of ⑯ ? ______________________________

20. What Det. # fits in the I.D. of ⑮ ? ______________________________

21. What is the purpose of the anchor pins? ______________________________

22. For Det. ③, what angle C'sink will be used? ______________________________

23. Calculate the minimum clearance between the .266 slot in ④ and the lever? ______________________________
____

TP-113, 113-1 & 113-2 (cont.)

24. The assembly drawing 113 was drawn first for functional layout, then each part was drawn in detail. Explain:

    A) Why the post was added? ______________________________

    ____

    B) Why (14) was shifted to the right in the sub-assembly view? ______________________________

    ____

    C) What was added to Det. (4) in Dwg. 113-2 making it look so different from the right side assembly view? ______________________________

    ____

TP-114

Answer the following questions:

1. In addition to the left section, front and top views, what other type of view is included in this drawing? ____________

2. What is the purpose of the X and Y reference lines? ____________

3. How many "B" holes are on this part, and what size are they? ____________

4. What is the general surface requirement for this part? ____________

5. How many holes are there in this part? ____________

6. How many tapped holes are called for? ____________

7. What does the symbol ⌒ over Datum [A] refer to? ____________

8. Are the arrows on cutting plane line "A-A" pointing in the proper direction for the section shown? ____________

9. How deep is the spotface for the "C" hole? ____________

10. Give tap drill sizes for the following:

	Tap Drill	Tap Drill Decimal Equiv.
a. #6-40 UNF	____________	____________
b. # 1/4-20 UNC	____________	____________
c. #6-32 UNC	____________	____________
d. #2-56 UNC	____________	____________

11. What dimensions are used to locate the X-X, X-Y, datum lines? ____________

TP-114 (cont.)

12. What are the vertical and horizontal locating dimensions for hole "F"? ______

13. Without considering holes, spotfaces, or C'bores, how many surfaces must be finish machined?

14. Will it be necessary to break edges on holes and corners of machined surfaces? ______

15. How far from hole A-1 will surface Ⓜ be milled? ______

16. Will any other feature be machined in surface Ⓜ? ______

17. In the left view, identify sides A and B. ______

18. How deep is C'bore A-4? ______

19. How would you produce the .016 radius at the bottom of the .8660 C'bores? ______

Chapter XI

ADVANCED BLUEPRINT READING

This chapter consists of shop prints and worksheets similar to those you worked on in Chapter X. These prints, although more complex than previous ones, are typical of the kind found in general machine and tool and die companies across the United States. Like any other skill, blueprint reading abilities improve with practice. Your transition to shop blueprints will be completed when you have finished the remainder of the NTMA prints.

TP-115

Answer the following questions:

1. Which two holes must be concentric?
2. Within what tolerance?
3. Explain "Line ream .2497-.2507" as noted on this drawing.
4. Describe the surface treatment of this part.
5. Locate the .311-.318 hole relative to the .4372-.4382 hole.
6. What are the overall dimensions of the crank?
7. What is the center distance between the .090-.087 and the .311-.318 holes?
8. Explain "tangential cuts" as noted on this drawing.
9. How will the .12 radius in the inner pockets be produced?
10. What type of fit would you expect between this part and Item #1?
11. What is the angle between the two arms?
12. What is the mean size of the .4372-.4382 Ø hole?
13. Would it be practical to drill the .311 hole before the .090 holes? Fully explain why.
14. The .312 R TYP applies in how many places?

TP-116

Answer the following questions:

1. Describe what this gage checks. ______________________________
2. What do the phantom lines represent? ______________________________
3. What do the initials P.F. mean? ______________________________
4. How will the surfaces on the left end of the gage be machined? ______________________________
5. Name the views in this drawing. ______________________________
6. Where would you find information on the size of the ball? ______________________________
7. What material is the part made from? ______________________________
8. What general heat treat processes are required on the gage? ______________________________
9. Why does the part test against the angular surface? ______________________________
10. Which operations would be used to make the .250 dia hole? ______________________________
11. What size drill will be used in this hole? ______________________________
12. What size piece of stock would the gage be made from? ______________________________
13. How will you know if the part is acceptable? ______________________________

TP-116 (cont.)

14. What is the reason for the 1/16 relief on the gage? ____________________

15. What is the distance from the top of the pad to the center of the hole? ____________________

16. What dimension would this be before hardening? ____________________

17. Assuming the shank of the ball measures exactly .2500, what size will the hole be ground to achieve a P.F.? ____________________

TP-117

Answer the following questions:

1. Calculate dimensions: (Y)________________ (Z)________________

2. What is the maximum size of circle (V) ? ________________

3. Will machining be necessary at (W) ? ________________

4. Should the surface at (U) be finish machined? ________________

5. What is the minimum size of (X) ? ________________

6. What is the vertical locating dimension for (X) ? ________________

7. What is the maximum width of the 1 1/2 TYP pad? ________________

8. What is the radius at: (S)? ________________ (T)? ________________

9. Referring to the 1/4-20 holes, what is the center distance between the holes? ________________
   Vertical ________________ Horizontal ________________

   What are the locating dimensions? Vertical ________________ Horizontal ________________

10. What tap drill will be used for the threads? ________________

11. Will it be necessary to C'sink the threaded holes? ________________
    To what approximate dia? ________________

12. Will a screw or bolt be used in the (2) .281 dia holes? ________________
    If so, what size screw? ________________
    What is the type of hole called? ________________
    Why is this hole spotfaced? ________________

    ____

13. The bottom view is shifted to the left. Is this a normal practice? ________________

    ____

    Why was this probably done? ________________

TP-117 (cont.)

14. In Sec. A-A, what is the thickness of the material shown directly above the .380-.383 dimension?

    ____

15. In the bottom view, the note | ⊙ A .002 TIR | refers to the concentricity of what dials? ______

    ____

16. Must the 1.939 and 1.687 dials be concentric within .002? ______________________________

17. What microfinish is required in hole | -A- | ? ______________________________

18. What will determine whether this part is sand cast or die cast? ______________________________

    ____

TP-118

Answer the following questions:

1. What is the tolerance of the .585 dimension? ____________

2. Where is the tolerance for Question 1 found? ____________

3. What is the common name of this part? ____________

4. Which machine tool is used to fabricate parts like this? ____________

5. What is the overall length and width of the part? ____________

6. In the enlarged view, what is the distance from the horizontal ℄ to the bottom line shown? ____________

7. What is the maximum diameter of the large hole? ____________

8. What is the smallest diameter permitted for the small hole? ____________

9. Why is an enlarged view shown? ____________

10. How many grooves in this part? ____________

11. What finish is required on the grooves? ____________

TP-119

Answer the following questions:

1. What is the material thickness? ____________________

2. What is the thinnest and thickest material permissible? ____________________

3. How many holes in the part? ____________________

4. What are the diameters of the holes? ____________________

5. In the .296 holes, what is the largest diameter hole permissible? ____________________

6. Explain how this information is obtained? ____________________

7. The .560 ± .010 dimension is typical of what other detail? ____________________

8. What is the radius of the slot that prevents the part from tearing while bending? ____________________

9. Give the overall dimensions of the part. ____________________

10. What are the locating dimensions for the .156 dia. hole?
    Vertical? ____________________ Horizontal? ____________________

11. What dimensions locate the .250 dia hole? ____________________

12. What is the length of the relief at the bottom of the front view? ____________________

13. In the view on the right, how will the angle be formed? ____________________

14. On a separate sheet of paper, carefully draw a 2x drawing of this part. Show what the part would look like before the ear was bent. Indicate bend line with a phantom line.

TP-120

Answer the following questions:

1. What are the overall dimensions of the part? ______________________________

2. What is the tooling hole? ______________________________

3. What is the largest diameter hole possible in the part? ______________________________

4. Assuming that the hole referred to in Question 3 is to be pierced out in a die, what diameter would you make the pierce punch? ______________________________

5. How much is the .080 hole offset from the V℄ in the front view? ______________________________

6. Allowing .150 for bends, what would the overall length and width of the part be before bending? ______________________________

7. At what point will the part be case hardened? ______________________________

If you were making one mock-up part for testing:

8. When would you machine the .3405 and .3725 dia holes? ______________________________

9. How would you form the "U" shape in the bottom view? ______________________________

10. On a milling machine, what size and type cutter would you use to form the 1/16 radii? ______________________________

11. Draw a free-hand sketch, close to actual size, of this part before the bends were made. (Use phantom lines to represent bend lines.)

TP-121

Answer the following questions:

1. What is the lower view called? ______
2. What does line X-X represent? ______
3. What purpose does line X-X serve? ______
4. What dimensions locate the .144 slots? ______
5. What is the tolerance on the width of the slots in the auxiliary view? ______
6. On which type of machine would a part like this be mass-produced? ______
7. What type H.T. will be used to achieve 60-65 RC? ______
8. What is the thickest material that can be used? ______
9. How many inside radius bends are there in this part? ______
10. At what angle is the auxiliary view formed? ______
11. What would be the length and width of the opening where the prongs are located? ______
12. What is the length of the lower slots? ______
13. Give the width of the material used. ______
14. What would be the basic distance between any two prongs? ______
15. Which direction should the grain run on this part? ______

TP-122

Answer the following questions:

1. What are the overall dimensions of the block? ______
2. How many individual parts does this gage consist of? ______
3. For what reason are all the details dimensioned on the assembly? ______
4. How many flush pins are there? ______
5. Why are the flush pins made to different lengths? ______
6. What material are they made from? Why? ______
7. What is the total tolerance that Detail (1) is checking? ______
8. What is the total tolerance that Detail (2) is checking? ______
9. What is the purpose of Detail (4)? ______
10. Why is the side view sectioned? ______
11. To what size would you make the .375 hole before hardening? ______
12. How would the .125 dia hole be machined? Why? ______
13. Where would the tool number and part number be stamped? ______

TP-122 (cont.)

14. Would this gage be used to check two different features on one part, or one feature on two different types of parts? ____

____

15. How would the $^3/_8$ x .070 flats in the flush pins be finish machined? ____________________

____

____

TP-123

Answer the following questions:

1. How many parts does the boring fixture consist of? ____________

2. Which parts have been removed from the right side view? ____________

   ____

3. Give the size of the four counterbore holes (using standard shop practices):
   a. Counterbore (diameter) ____________, depth ____________
   b. Drill size (diameter) ____________, depth ____________

4. How are surfaces to be ground noted? ____________

   ____

5. How parallel and square must the ground surfaces be to the center-line? ____________

   ____

6. What does the note "LAP PROTECTED CENTERS" mean? ____________

   ____

7. What is the largest diameter in Part (1)? ____________

8. What is the overall length of Part (1)? ____________

9. What does "1 1/8 TYP. CENT." mean? ____________

   ____

10. How many tapped holes does this fixture have? ____________
    What is/are their size(s)? ____________

11. What is the diameter of the bolt circle? ____________

12. Why is the 3/8-16 tapped hole C'drilled to 1/2 dia? ____________

    ____

13. What is the purpose of the chamfer on the part feature with the 2.7510-2.7508 dimension?

    ____

TP-123 (cont.)

14. What is the length of the 5 3/8 dia? ______

15. What is the function of the four C'bored holes in Part (3)? ______

16. How deep is the tapped hole? ______

17. What is the function of Det. # (3)? ______

18. What is the dia and depth of the C'bore in Det. # (3)? ______

19. Why was this C'bore added to Det. # (3)? ______

20. What function do the 13/16 holes serve? ______

21. What is the purpose of the .750 x 3/8 long dia? ______

TP-124

Answer the following questions:

1. What dimensions locate the holes for Det. (3)?
   Vertical? ______________ Horizontal? ______________

2. What is the purpose of the two C'bored holes? ______________

3. Calculate the horizontal center distance between the C'bored holes. ______________

4. What kind of material is the base made from? ______________

5. How are Parts (3), (4) and (5) secured to the base? ______________

6. What kind of machining operations will be performed on the top of Part (4)? ______________

7. Why is Det. (2) altered to an oval point? ______________

8. How far will Part (3) protrude above the surface of the fixture? ______________

9. Is it necessary to grind surfaces (A) (B) (C) and (D) ? ______________

10. What surfaces on Det. (1) must be ground? ______________

11. What is the purpose of Det. (2)? ______________

12. Which of the following methods is best suited to locating the positions of all the holes on the fixture?
    a. Jig Bore
    b. Milling machine
    c. Layout with surface gage and scale
    d. Layout with vernier height gage
    e. None of the above

TP-124 (cont.)

13. Give the dimensions of the base plate. ______________________________

____

14. What are the total number of parts in this fixture? ______________________________

____

15. What is the function of Part (6)? ______________________________

____

16. What is the minimum and maximum thickness of Part (8)? ______________________________

____

17. Describe the surface hardening technique for Part (8)? ______________________________

____

18. What size is the socket head cap screw? ______________________________
What size tap drill should be used here? ______________________________

19. What threads will be used in Det. (4)? ______________________________

20. What is the tolerance of the .625 dia P.F. holes. How is this determined? ______________________________

____

____

21. To what dimensions does the ± .005 tolerance apply? ______________________________

22. If Det. (2) were rotated one complete turn clockwise, how far will it travel into the base?

____

TP-125

Answer the following questions:

1. What diameter of Det. (3) fits into the C'bore of the #1326 faceplate? ______

2. What threads are on the ℄ of the #1326 faceplate? ______

Why is the C'washer:

3. Slotted; ______

4. C'bored? ______

5. C'bored on both sides? ______

6. Knurled? ______

In Det. 3 :

7. On what surfaces will the workpiece locate? ______

8. Why is this Det. hardened? ______

9. Are there threads in the $^{15}/_{32}$ hole? ______

10. What is the function of the $^{1}/_{8} \times 15^{\circ}$ chamfer? ______

11. Why is the $2^{1}/_{8}$ ⌀ undercut? ______

12. What is the purpose of Det. (5)? ______

13. Will Det. (2) be stamped after hardening? ______

14. How much of Det. (3) will extend into the #1326 faceplate? ______

TP-125 (cont.)

15. In the 1/4-20 tap note in Det. (1), what is meant by "LOC FROM DET (3)"? ______________
____
____

16. Allowing .007 stock on a side for grinding, how deep will you C'bore Det. (2) before hardening?
____

17. Would it be best to C'bore (2) before or after the slot? ______________________
Why? ____

18. What size screws hold Det. (1) to the faceplate? ______________________________
____

19. On what type of grinder will (C) be ground and how? __________________________
____

TP-126

Answer the following questions:

1. Why are the .375 holes spotfaced? ______________________

2. How deep are the spotfaces? ______________________

3. What is the maximum and minimum depth of the .562 Ø C'bore?
   Maximum: ____________ Minimum: ____________

4. What approximate dia pilot will be used with this C'bore? ______________________

5. What would be the low limits of the $1\frac{1}{8}$ and $1\frac{5}{32}$ dimensions in the top view, center pad?
   $1\frac{1}{8}$ Lo = ____________ $1\frac{5}{32}$ Lo = ____________

6. What are the center distances between the .190 dia holes?
   Vertical: ____________ Horizontal: ____________

7. What is the vertical center distance between the .375 dia holes? ______________________

8. What is the vertical locating dimension for the .375 holes? ______________________

9. Would the (3) optional pads serve a purpose, other than to use them for your setups? ______________________

10. Without considering the optional pads, what surface would you machine first? ______________________

11. In the .6245 dia holes; if the two outer holes were in line and one inner hole was +.001, the other -.001, would the hole alignment be acceptable? ______________________

12. Will the top surface in the top view be finish machined? ______________________

13. What is the length and width of the back side of the casting? ______________________

TP-127

Answer the following questions:

1. What is another term for welded construction? ____________________

2. How many pieces are in this assembly? ____________________

3. What type of steel is the part made from? ____________________

4. What does "stress relieve" mean? ____________________

5. How many holes are drilled in the part? ____________________

6. What are the maximum and minimum acceptable values of the 22° angle? ____________________

7. Name the three views starting from the left side of print. ____________________

8. Should stress relieving occur before or after finish machining? ____________________

9. At what angle is (D) machined? ____________________

10. What does the ~ mark under some dimensions mean? ____________________

TP-128

Answer the following questions:

1. How would this part be produced in a large quantity? __________

2. What thickness is left between the two .030 R in the top view? __________

3. What general heat treating process will be used on this part? __________

4. How many bends are on this part? __________

5. Calculate the "X" dimension. __________

6. What does the 1/4 dimension in the right side view represent? __________

7. In making a model part, how would the .127 holes be machined? __________

8. Referring to Question 7, how would the U-bend in the top view be formed? __________

9. Which locating points would you use to form the U-bend? __________

10. Locate the .216 radius. __________

11. Locate the .078 radius. __________

12. What size piercing punch would you make to pierce the $.127^{+.002}_{-.000}$ holes? Why? __________

TP-129

Answer the following questions:

1. How deep is the 1/2 wide slot? ______

2. How would you machine this slot? ______

3. What machine tool would be used to rough machine the 1 1/8 long neck? ______

4. How would you hold this part in the above-named machine? ______

5. Would it be pracitcal to use 1 1/4 x 2" C.R.S. barstock for this part? ______

6. If you used 1 1/4 x 2" stock, would it be practical to lay out diagonal lines for the center drilling? ______

7. No specially purchased tooling is available to produce the ring grooves. How might you do this machining? ______

8. In Sec. A-A, what is the distance from the right side of the groove to the far end of the hole? ______

9. In how many places does Sec. A-A apply? ______

10. What tap drill will be used on this part? ______

11. What is the *mean* dia of the .8745 to .8750 dia neck? ______

12. Would it be permissible to layout the .375 ℄ holes? ______
    Why? ______

TP-130

Answer the following questions:

1. How many parts is this weldment made from? ____________________
   Name them: ____________________

2. What material is used in each of the details? ____________________
   ____

3. What type of line is used to show the shape of the part? ____________________
   ____

4. How many large slots are in the part? ____________________

5. What are the lengths of the slots? ____________________

6. Would it be practical to layout and cut the six slots before rolling and welding? ____________________
   ____

7. Calculate the length and width of the skin.
   Length: ____________________ Width: ____________________

8. What is the outside diameter of Part (3)? ____________________

9. Why is the end flange shown in a separate section view? ____________________
   ____

10. What is the total tolerance on diameter X? ____________________

11. How thick is Part (2)? ____________________

12. What is the overall length of the part? ____________________

13. What finish is required on the machined surfaces? ____________________
    ____

TP-131

Answer the following questions:

1. Det. (1) will require 1 piece of 1" x $2^{1}/_{8}$" x $14^{1}/_{8}$" alum. 6061-T6 stock, since it must machined on both ends and surface [B]. Make a bill of material to order the remaining 8 details.

Stock and cutoff sizes

(2) ______________________________
(3) ______________________________
(4) ______________________________
(5) ______________________________
(6) ______________________________
(7) ______________________________
(8) ______________________________
(9) ______________________________

2. What heat treatment is required? ______________________________

3. Why are some dimensions missing from the .187 drilled holes? ______________________________
____

4. Will it be necessary to know the exact size of Det. 81 , in order to machine the .750 dia. hole?
____

5. What are the internal sizes of Det. (8)? ______________________________

6. How many surfaces will be machined after welding? ______________________________
____

7. At the bottom of the right view, what does the circular hidden line represent? ______________________________
____

8. Why is it necessary to stress relieve this part before machining? ______________________________
____

9. To what size will you countersink the $^{3}/_{8}$-24 threads? ______________________________

10. Draw Section A-A on the print above title block.

TP-132

Answer the following questions:

1. What dimensions locate the two 3/8 holes in the base? ______________________________

2. What body drill will be used in the Det. (5)? ______________________________

3. What material is the base made from? ______________________________

4. What is the purpose of the 3/8 dia holes? ______________________________

5. Would a surface plate be used in conjunction with this inspection operation? ______________________________

6. What is the purpose of Part (8)? ______________________________

7. Why is Part (8) hardened? ______________________________

8. What are the overall dimensions of the base? ______________________________

9. How can the undercuts in the bracket be produced? ______________________________

10. Describe the operations necessary, and tooling required, to machine the elongated slot in (5).
____

11. What is the reason for this slot? ______________________________

12. When transfer punching Det. (5), what dimension is used to locate that detail? ______________________________

TP-132 (cont.)

13. Based on common stock allowances, what approx. size will you leave the following dimensions for finish machining:

    In the Base:

    3.437 =

    $2^3/_8$ =

    $^7/_8$ dim at bottom =

    In the bracket:

    .500 =

    $^7/_8$ =

    $^5/_{32}$ =

    In the set block:

    3.437 =

    .478 =

    $1^3/_4$ =

    $^1/_2$ =

14. In your opinion, are the undercuts necessary in the set block? ____________________

    ____

    ____

TP-133

Answer the following questions:

1. How many holes will be drilled into the workpiece that this jig is made for? ____________________
2. In the top view, what is the purpose of the large relief below Det. (14) &(9)? ____________________
3. What is the function of Det. (13) ? ____________________
4. What precaution should be taken when reaming the 3/8 Ø x 1/4 DP hole in Det. (9)? ____________________
5. What machine should be used to put in the two 5/16 holes for Det. (10) ? ____________________
6. Should these 2 holes be machined after Det. (9) is assembled? ____________________
7. List the dimensions used to locate the 2 bushing holes:
   1) ______________ 2) ______________ 3) ______________
8. Should the rest buttons be ground before or after they are press fitted? ____________________
9. What function does Det. (18) perform? ____________________
10. What is the most productive way to form the full R in (5)? ____________________
11. Why was the gusset added? ____________________
12. Calculate the angle between the 2 Van Kuren pins in the front view. ____________________

TP-133 (cont.)

13. Why was the .81 x 1.31 relief milled into Det. (9)? ______________________________

____

14. Should the top of the gusset be machined flush with the top of the "T" section?

____

Why? ______________ How? ______________

15. What detail is used to maintain the accurate alignment of the Det. (14) ? ______________

____

16. What Det. # holds Det. (14) to Det. (1)? ______________________________

17. Describe the part that is used to hold the workpiece against Det. (14) while drilling? ______________

____

18. Assuming Det. (13) measures exactly .1875, what size would you make its mating hole in (9)?

____

19. The fixture is now assembled and ready to bore the bushing holes. Will it be necessary to remove Det. (9) to locate the position of these holes? Explain. ______________________________

____

____

TP-134

Answer the following questions:

1. In the right view, what is the dimension across the flats of the locator plug? ____________________

2. What is the purpose of the hooked portion of the elongated slot in (7)? ____________________

3. On what dia, and against what surface, will the workpiece locate?
   Dia: ____________________ Surface: ____________________

4. What is the purpose of Det. (8)? ____________________

5. What does "TO SUIT" mean in Det. (8)? ____________________

6. What is the function of Det. (2)? ____________________

7. A) How would you inspect the parallism between Det. (19) and (1)? ____________________

   B) What T.I.R. is required. ____________________

   C) In what distance approx.? ____________________

8. A) Would it be helpful to put centers on both ends of (19) ? ____________________

   B) Why? ____________________

   C) Would centers be permissible? ____________________

9. What would the depth be of the 60° "V" in the right view? ____________________

10. Compare the finish stock sizes in the data block to those sizes on drawing. Are there any *major* discrepancies? ____________________

11. In Det. (13) is .083 a standard drill blank size? ____________________

TP-134 (cont.)

12. What standard size could be used for (13) ? ______________________
Why? ______________________

Note: Det. (14) is a standard knob which has a slot milled in it. This slot removes over 50% of the threads. After 2 or 3 turns it can be tilted and pulled off quickly, saving loading and unloading time on each piece.

13. It is likely that a lot of cutting pressure will result in this milling operation? ______________________
____

14. Assuming a 3″ dia by 60° form cutter is being used, on a horizontal miller, which way will the cutting pressure be directed:
A) toward Det. (15) , or B) toward the "T" section? ______________________

15. What is the purpose of the 60° groove on the top of the "T" section? ______________________
____

16. What is the purpose of the 3/4 hole in the base? ______________________
____

TP-135

Answer the following questions:

1. How many holes are on the 9.500 dia B.C.? ______

2. What is the purpose of the undercut? ______

______

3. How many clamp assemblies are on this fixture? ______

4. How will the surfaces marked "G" be machined? ______

5. What do the phantom lines on this drawing represent? ______

______

6. How many 5/16-18 tapped holes are required? ______

7. What type of part is called for at (8)? ______

8. What function does (8) serve? ______

______

9. Give a reason why the 4.7500 bore has a $^{+.0005}_{-.0000}$ tolerance. ______

______

10. How many 11/32 holes are called for, and what is their purpose? ______

______

11. Does the "Section A-A" correspond exactly with the cutting plane line? ______

______

______

12. What is the function of Part (6)? ______

______

13. How deep is the 12.500 Ø C'bore? ______

14. On what surfaces will the workpiece locate? ______

______

TP-135 (cont.)

15. If the keyway for Part (6) were cut as shown, specify the type and size of cutter needed. ________
____
____

16. Why was the clamp altered to accept Det. (2)? ______________________________
____

17. At balloon (3) what is meant by "LOCATE FROM DET. (2)"? ____________________
____

TP-136

Answer the following questions:

1. What does the symbol ⊥A .0005 mean? ______________________________
____

2. Locate Datum [A]. ______________________________

3. Which two surfaces must be square to the .3125-.3127 dia hole? ______________________________
____

4. Explain what the .001 squareness tolerance of the face located by the .239 dimension means.
____
____

5. Explain the note "SPOTFACE .109 DIA TO CLEAN." ______________________________
____

6. Why are some of the dimensions left off this drawing? ______________________________
____

7. Why are the finish marks left off this drawing? ______________________________
____

8. Interpret the note ".015 DIA THRU & C'SINK 90° TO .03 DIA ON A .400 R." ______________________________
____
____

9. Explain the symbol ⊙ A .001. ______________________________
____

10. Explain the note ".750 DIA. B.C." ______________________________
____

TP-137

Answer the following questions:

1. What tap drills will be necessary to manufacture this jig? ________________

2. What dimensions are used to locate Det. (2)? ________________

3. What are the locating dimensions for the angled swivel pad clamps? ________________

4. How will you produce the undercut in Det. (2)? ________________

5. What is the purpose of the 3/16 drilled holes on the bottom? ________________

6. What detail holds the locator on? ________________

7. No parts on this drill jig are hardened. Why? ________________

8. Are the lock screws a type of shoulder screw? ________________

9. When preparing to bore the bushing holes, would it be advisable to use a 3/16 drill first?
________________ Why? ________________

10. Is line reaming necessary on this part? ________________
Where? ________________
What size reamers? ________________

11. Give a reason for the use of only 2 slip bushings – of 2 different sizes – in this four hole workpiece?
____

TP-138

Note: Suggested reference: *Basic Diemaking,* National Tooling and Machining Association.

Answer the following questions:

1. List the parts of the die set
   A. ______
   B. ______
   C. ______
   D. ______
   E. ______

2. What operation will this die perform? ______

3. How is the punch held onto the upper shoe? ______
   ______

4. Describe the part that prevents shifting of the nests. ______
   ______

5. Will it be necessary to have samples of the part available to position the nests? ______
   Why? ______
   ______

6. What do the circular phantom lines represent? ______
   Why are they shown? ______
   ______

7. Why are there no locating dimensions or center distances shown for the holes in the punch?
   ______

8. Will Det. (11) be press fitted in both the punch and shoe? ______

9. Besides showing the width of the die block, how will the 3″ dimension be used? ______
   ______

10. Will the punch be placed on the ℄ of the ram? ______
    How do you know this? ______
    ______

TP-138 (cont.)

11. What locating dimension will you use to position the dowel holes in Det. (3)? ________________

____

12. The hole for Det. (11) is drilled thru. Why? ________________________________

____

13. Assuming the punch is 90°0′0″, will the part after bending be: more than 90°0′0″, less than 90°0′0″, or exactly 90°?

____

TP-139

Answer the following questions:

1. Will the surfaces indicated below be finish machined?
   (A) __________ (B) __________ (C) __________ (E) __________ (F) __________

2. How deep will you tap drill (D) ? __________

3. List the tap drills that can be used for (D) . __________
   ____

4. Would using the smallest tap drill produce a better class of fit in (D) ? __________
   What would change by using the smallest tap drill? __________
   ____

5. Calculate the wall thickness (G) . __________

6. Are the surfaces that are held parallel within .002, in the same plane? __________
   ____

7. What is the vertical center distance between the two holes shown .094 apart in the section view?
   ____

8. Is the .552 dia a thru hole or C'bore? __________

9. Would the tap for the 10-32 NF3 thread be smaller or larger, on the pitch dia, than a tap used for Class 2? __________

10. What special preparation will be necessary on the points of drills used on this part?
    ____
    ____

TP-140

Answer the following questions:

1. Which of the .500 holes is the most important in the construction of the slots and location of the other .500 holes in Det. (1)? ______________________

2. What is the width and depth of the slots in (1)? ______________________

3. Will plates (3) and (4) be mounted on Det. (1) as sub-templates? ______________________

4. Is it likely that (3) and (4) will be mounted on (1) at the same time? ______________________
   Why? ______________________

5. Is it possible that either (3) or (4) be used in an upside down position from that position shown on print? ______________________
   Why? ______________________

6. Calculate dim "X" and tolerance. ______________________

7. Calculate the angular relationship between (E) and (C). ______________________

8. What is the center distance between (C) and (E)? ______________________

9. Will another diamond pin be used in (E)? ______________________
   Why? ______________________

10. The tolerance for four place decimals is not shown. Did the draftsman forget to list this tolerance?
    ____

11. Would you be inclined to put these holes in on a jig borer rather than on a Bridgeport?
    ____

TP-141

Answer the following questions:

1. Does the $1^3/4$ dimension in the auxiliary view locate: the ℄ of the diamond pin from the edge of block or the edge of the block from the ℄ of the diamond pin? ______

2. Would it be practical to finish bore the diamond pin hole in the block before it is line reamed and screwed to the base? ______
   Explain. ______

3. Why is only *one* dowel used to hold the alignment of the block? ______

4. How will these construction holes be used in your machining set-ups? ______

5. What inspection equipment will be used when checking the parts that will be mounted on this gage? ______

6. What will the minimum I.D. of part #473-R be? ______

7. How far from the ℄ of the base is the chamfered edge of (8)? ______

8. Does the 15° REF. correspond with the construction hole locating dimensions? ______

TP-142

Answer the following questions:

1. How many pieces make up this weldment? ______

2. How many surfaces are machined after welding? ______

3. What material is this weldment made from? ______

4. What do the two wavy lines in the middle view mean? ______
______

5. What does S.F. mean in the note in the upper righthand corner? ______
______

6. Why is a 1/4-20 thread used in conjunction with the diamond pin hole? ______
______

7. Is there any heat treatment on this part? ______

8. What do the phantom lines represent on the bottom view? ______
______

9. What do the xxx's mean? ______
______

10. What is the purpose of the diamond-shaped pin? ______
______

11. Assume a 10″ center distance between construction balls X and Y. Assume balls X and Z are equidistant from the end of P/N 1595 and 7″ apart center to center. At what angle is the front view set? ______
At what angle is the right view set? ______
Give answers in degrees, minutes *and* seconds.

TP-143

Answer the following questions:

1. Calculate dimension X in section G-G. ______

2. Is the smallest diameter on the ℄ of this part to be finish machined? ______

3. What is the size of the diameter that surface [B] must be squared to within .001? ______

4. Based on given angle, how deep will you C'sink the .875 dia hole, to achieve the .940 diameter? ______

5. Which combination of circled letters are in the same plane? ______
   AB AC AD BC BD CD

6. What size and type cutter will be used to machine the groove with the 2.199R.? ______

7. Design and dimension a cutter to spot face the five holes in the right view.

8. Which of the five .342 holes is closest to the center of the part in the left view? (Do not guess.) ______

Chapter XII

BLUEPRINT TERMS AND ABBREVIATIONS

Listed below are the common terms and abbreviations used in conjunction with most blueprints. Accompanying each is a definition and, in some cases, a diagram is included for clarity. The terms are in alphabetical order. The third column gives the accepted abbreviation that appears on blueprints.

Term	*Definition*	*Abbrev.*
Addendum	–The height of a gear tooth from the pitch circle to the outside diameter of the gear.	Add.
Adjacent	–"Next to"; an angle that shares a common leg with another angle.	Adj.
Aluminum	–A lightweight, white, soft, nonferrous metal produced from bauxite ore.	Alum.
American Iron and Steel Institute	–Professional organization of engineers who are responsible for research and standards in the steel and iron industry.	AISI
American National Standard Pipe Threads	–60 deg thread form cut straight or (more commonly) tapered at 3/4 in./ft. Pipe is used to conduct fluid or gas.	ANPT
Angle	–A geometric figure formed by two lines meeting at a point.	Ang.
Approximately	–An estimated or general statement about an object, i.e., nominal size, etc.	Approx.
Assembly	–A mechanism consisting of two or more parts placed in proper location.	Assy.
Auxiliary	–An orthographic view not contained in any of the six regular planes of projection, but constructed from one or more of them.	Aux.

Term	*Definition*	*Abbrev.*
Bearing	—An assembly of rollers, or balls and races, used to support linear and/or rotary motion. Bearing metal is a soft metal (i.e., brass, bronze, babbit, etc.).	Brg.
Blueprint	—A photographic copy of an engineering drawing; a graphic communication from a designer or engineer that tells a mechanic what an object looks like.	B/P or BP
Body	—The largest diameter to be found on a screw. Also, the main portion of an object.	
Bolt Circle	—Circular centerline for any group of part features, usually bolt holes.	BC
Bore	—The inside diameter of a cylinder. Bore size is usually designated by the length of the diameter.	
Boss	—A raised, machined surface that adds strength, facilitates assembly, provides for fastening, etc.	
Brass	—A non-ferrous metal that is an alloy of copper and zinc. Color ranges from yellow to red.	Br.
Brinell	—Upsetting a metallic surface with repeated blows, causing stress and hardness by compressing the microstructure of the metal.	
Brinell Hardness Tester	—Materials testing device that determines a metal's resistance to indentation or cutting. Usually, a small steel ball is pressed into the metal surface with a standard force. The spherical surface area of the resulting indentation is measured, and divided into the load. The results are expressed as a Brinnell Number.	
Broach	—A multiple-tooth bar-like cutting tool. Also, a machining operation that utilizes progressively sized and shaped cutters mounted consecutively, that gives a desired surface or contour.	
Broken-Out Section	—The region of an object graphically removed to show inner detail. Bounded by an irregular break line.	
Bronze	—A non-ferrous, yellow metal alloy composed mainly of copper and tin.	Brnz.
Cadmium	—A blue-white metallic element used as protective plating and in bearing metal.	Cad.

Term	*Definition*	*Abbrev.*
Capacity	– Indicates size by volume, weight, or both. Capacity may be given in gallons, cubic feet, pounds, etc., whichever is most descriptive according to the use of the container.	Capy.
Case Hardness (Case Harden)	– The hardness of the outer layer (case) of a part. To harden only the outer surface of a part by such methods as flame hardening, carburizing, nitriding, etc.	CH
Casting	– A part produced when molten metal is poured into a pre-formed cavity, and allowed to solidify before removing.	Cstg.
Cast Iron	– A brittle, ferrous metal alloy containing large quantities of carbon that is cast into a shape.	CI
Celsius Temperature Scale	– The metric temperature measurement scale, in which 0°C is the freezing point of water, and 100°C is the boiling point of water. Formerly called Centigrade.	°C
Centerline or Center Line	– The imaginary horizontal or vertical line passing through the center of a part feature, and extending infinitely in both directions.	CL or ℄
Center to Center	– Any reference to the imaginary line running between and intersecting the centers of two part features.	C to C
Chamfer	– A bevel or angle cut across the edge of a part to give a finished look, or to remove sharp edges.	Cham
Circumference	– The perimeter of a circle. Also the distance around a circular part. The length of the periphery of a circle or circular part.	Circum.
Clockwise	– Rotation in the same direction as the hands of a clock.	
Cold-Rolled Steel	– Barstock which has been rolled and shaped at room temperature. Generally has a smoother surface finish and more accurate rough dimensional size (±.002) than hot-rolled steel.	CRS
Concentricity	– The state of a shaft wherein the position of consecutive diameters lie along the same axis.	⦿
Core	– The part of a casting mold that forms internal holes in a casting. When the casting is cool, the core is broken out, leaving a cavity in the casting.	
Core Hardness	– The degree of hardness of the core of a steel part that is refined after carburizing.	

Term	*Definition*	*Abbrev.*
Cosine	–A trigonometric function used to solve for an unknown leg of a triangle when the side adjacent to the angle (SA) and the hypotenuse are known. $\text{Cos } A = \frac{\text{SA}}{\text{Hyp.}}$	Cos
Cotangent	–Trigonometric equation which states that the value of an angle in a triangle is equal to the side adjacent (SA) divided by the side opposite (SD). $\text{Cot } A = \frac{\text{SA}}{\text{SD}}$	Cot
Counterbore	–A flat-bottomed enlargement of the mouth of a cylindrical bore used to set the head of fastener below the surface of the work. Also, to enlarge a borehole by means of a counterbore.	C'Bore
Counter-Clockwise	–Rotary motion in the direction opposite that of the hands of a clock.	CCW
Counterdrill	–A taper-bottomed enlargement of the mouth of a hole used to set the head of a fastener below the surface of the work.	C'Drill
Countersink	–A bevel or flared depression around the edge of a hole used to set the head of a flathead screw below the surface of the work.	C'Sink
Crankcase	–Crankshaft housing, i.e., engine block.	C-Case
Cylinder	–A geometric figure having a circular cross section.	Cyl.
Dedendum	–The distance from the pitch circle to the base of a gear tooth.	
Degree	–Unit of temperature measurement, i.e., degrees Fahrenheit or degrees Celsius. Also, unit of angle measurement = $^1/_{360}$ part of a circle.	° or Deg
Detail	–Special drawing of an object, or a portion of an object, that requires more information than is normally included on a working drawing. The detail may be isolated or enlarged to attract attention to some important aspect or feature.	
Diagonal	–In a regular square or rectangle, the line drawn between opposite corners; slanting, oblique.	Diag.
Diameter	–A line segment passing through the center of a circle, and whose end points lie on the circle.	Dia or D

Term	*Definition*	*Abbrev.*
Diametral Pitch	—Ratio of the number of teeth on a gear to the pitch diameter; equals the number of gear teeth per inch of pitch diameter.	DP
Die	—One of a pair of hardened metal blocks for forming, impressing, or cutting out a desired shape. Also, a tool sometimes used to cut external screw threads (thread-cutting die).	
Die Casting	—A very accurate and smooth casting made by pouring a molten alloy (usually under pressure) into a metal mold or die. Distinguished from a casting made in sand.	
Dimension	—Numerical value expressed in appropriate units of measure (inches, fractional inches, decimal inches, millimeters, degrees, etc).	Dim.
Direction	—Movement toward a special location.	Dir.
Drawing	—A graphic representation (sketch, blueprint, etc.) of an object. A collection of straight lines, curves, and dimensions that shows the shape and size of an object.	Dwg.
Drilled Hole	—A hole created or enlarged by a drill bit. Distinguished from a bored or reamed hole.	
Drill Rod	—Carbon tool steel purchased in round or square cross section in lengths of 3 ft and 1/8 to 1 in. in diameter. Used to make cutters and pins of various types.	DR
External	—The area or surface on the outside of an object.	Ext.
Face	—To machine a flat surface. A cut taken on a lathe to square the end of stock with its axis of rotation. Also, working surface of a gear tooth above the pitch circle.	
Fahrenheit Temperature Scale	—The temperature measurement scale in which 32°F is the freezing point of water, and 212°F is the boiling point of water.	°F
Figure	—An illustration, a numeral, or a geometric shape.	Fig.
Fillet	—A curved or rounded surface between two intersecting surfaces. The radius cast in the joint between intersecting surfaces of a casting.	Fill.

Term	*Definition*	*Abbrev.*
Fillister Head	—A type of screw head having long, straight sides (much like a socket-head screw) and a dome-shaped top with a single slot.	Fill. Hd.
Fin	—A thin metal projection rib found especially on cylinder heads, and used to dissipate heat.	
Finish	—Any final surface preparation on a part that protects it, makes it function more efficiently, and/or improves its appearance. Also used to designate surface quality and surface texture.	Fin. or √
Flange	—A projecting rim or rib used for strength, guidance, or as a means of attaching to another object.	
Flat Head	—82 deg tapered head fastener designed to set below the surface of the work.	Fl. Hd.
Forging	—The process by which metal is shaped by compressive force (hammer, press, rolls, etc.). May be accomplished while metal is hot or cold, depending upon the manufacturing process.	
Gage	—A standard of reference for size or shape, i.e., gage block, thread gage, etc.	GA
Gear	—A wheel or disk, having teeth around its periphery, used to interlock with other gears to transmit motion.	
Grind	—To remove material from, or reduce the size of, a workpiece by contact with an abrasive wheel. To finish or polish a flat or curved surface with an abrasive wheel. To sharpen with an abrasive wheel.	G
Harden	—The most important of the heat treating processes, hardening increases the tensile properties of metals. To harden metals by heating and cooling at predetermined rates.	Hdn.
Height	—A vertical dimension.	Ht.
Hexagon	—A six-sided polygon.	Hex.
Horizontal	—Parallel to the plane of the horizon.	Hor.
Hypotenuse	—In a right triangle, the longest side opposite the 90 deg angle.	Hyp.

Term	*Definition*	*Abbrev.*
Inch (one)	– 1/12 of 1 foot, equal to 2.54 cm.	In. or "
Included	– Found within or between i.e., an included angle or side.	Incl.
Inside Diameter	– The length of a line drawn through the center of a cylinder or sphere, terminating at the inside circumference on each end.	ID
Inspection	– Careful examination of an object by a qualified person to determine whether or not the object conforms to applicable specifications or quality standards, such as the dimensional accuracy of a part. May be an interim or final check.	Insp.
Internal	– The surface or area inside an object.	Int.
Junction	– The point of intersection of two or more lines, planes, axes, or features.	Jct.
Keyway	– In a mechanical power transmission system, the pocket in the driven element that provides a driving surface for the key. A groove or channel in a shaft or in the hole of a gear or pulley that fits a key to prevent joint slippage.	
Left-hand	– Oriented to the viewer's left hand. A left-hand screw thread advances into a material when turned counter-clockwise.	LH
Linear	– Of or relating to a line. In a straight line, i.e., linear motion. Having a single dimension.	Lin.
Line Ream	– To ream holes in-line, or parallel to each other.	
Lug	– A projection or head on a metal part that serves as a cap, handle, support, or fitting connection. A projection from a casting of irregular shape.	
Machine(ing)	– To perform various cutting or grinding operations on a workpiece.	Mach.
Machine Steel	– Free-machining, general-purpose plain carbon steel with a 0.2 to 0.3% carbon content.	MS
Magnaflux	– Trade name for the non-destructive magnetic particle materials testing method, which is used only on magnetic steels to identify crackes. The term has been replaced with *magnetic crack testing* and *magnetic particle inspection*.	M

Term	*Definition*	*Abbrev.*
Magnesium	—A silver-white lightweight, malleable, ductile metal.	Mag. or Mg
Manufacturing	—That portion of the industrial world involved in producing the goods that our society uses day-to-day.	Mfg.
Material	—The substance from which an object is made.	Mat'l
Maximum	—The greatest quantity, amount, or degree. The highest possible value.	Max.
Mechanical	—Of, pertaining to, or concerned with machinery or tools.	Mech.
Milling	—To machine metal surfaces by forcing a part past a rotating cutter.	
Minimum	—The least quantity, amount, or degree. The least possible value.	Min.
Minute	—A unit of time equal to $^1/_{60}$ of an hour or 60 seconds. Also, a unit of angle measurement equal to $^1/_{60}$ of 1 degree.	Min.
National Coarse	—A screw thread designation for a coarse thread.	NC
National Extra-Fine	—A screw thread designation for an extra-fine thread.	NEF
National Fine	—A screw thread designation for a fine thread.	NF
National Pipe Threads	—American National Standard Taper Pipe Threads: a 60 deg thread form for pipe.	N.P.T.
Number	—A symbol or word expressing a quantity.	No. or #
Numeral	—A symbol that expresses or names a number: 1, 2, 3, etc.	
Opposite	—A part feature in line with another feature but separated by space.	Opp.
Out-of-Round	—Not round. Egg shaped or elliptical. Not turning on a geometrical axis.	
Outside Diameter	—The length of a line drawn through the center of a cylinder or sphere, terminating at the outside circumference on each end.	OD
Overall	—A general statement pertaining to an entire object.	OA

Term	*Definition*	*Abbrev.*
Pad	–A projection of excess metal on a casting, forging, or welded part. Usually irregular in shape.	
P & 1	–Primer coat plus one coat of paint.	
Parallel	–A situation wherein two or more lines or planes are at a fixed distance from one another, never converging or diverging.	
Pattern	–The exact replica of a finished casting, except that it is slightly larger to allow for shrinkage. Made from wood or metal, the pattern is set in sand or similar material to form the cavity of the mold.	Patt.
Pitch	–In a screw thread, the distance from a point on a thread to the same point on an adjacent thread.	p
Pitch Diameter	–The diameter of an imaginary circle on a gear that separates the addendum and dedendum of the tooth. RPM calculations for gears are based on this circle, because (theoretically) the pitch circles of mating gears are tangent.	PD
Radius	–Equal to ½ the diameter of a circle. The distance from the center of a circle or sphere to any point on the periphery.	R or Rad.
Reamed Hole	–Any previously-drilled or -bored hole that has been sized, enlarged, or smoothed by a cutting tool called a *reamer*.	
Reference Dimension	–A dimension without a tolerance used for informational purposes only, and does not govern machining operations in any way. A distance from a point or plane used to locate a part feature.	REF.
Revised or Revision	–Any formal changes made to a blueprint.	Rev.
Revolutions Per Minute	–The number of times in one minute that a body spins 360 deg on its axis.	rpm
Right Hand	–Oriented to the viewer's right hand. A right-hand screw thread advances into a material when turned clockwise.	RH
Root Diameter	–The diameter of a screw taken at the bottom of the grooves. Also, the diameter of a gear taken at the root circle, or base, of the teeth.	RD
Round Head	–Shape of a common screw fastener's head representing one-half of a sphere.	Rnd. Hd.

Term	*Definition*	*Abbrev.*
Screw	—A cylinder with a helical groove cut into its surface.	
Section or Sectional View	—The graphic removal of a portion of an object to reveal internal lines and surfaces.	Sec.
Sine	—A trigonometric function showing the relationship between the value of an acute angle in a right triangle, and the side opposite that angle and the hypotenuse. $\text{Sin } A = \dfrac{\text{Leg opposite A}}{\text{Hypotenuse}}$	Sin
Society of Automotive Engineers	—Organization that promotes all aspects of the design, construction, and use of self-propelled mechanisms, prime movers, their components, and related equipment.	SAE
Socket Head	—The head of a screw fastener in which a hexagonal-shaped pocket is recessed. Tightened with an Allen wrench or hexagonal key. Standard for capscrews.	Skt. Hd.
Specification(s)	—An organized list of basic requirements for materials, product composition, and dimensions. The information on a blueprint relating to an object's size, surface quality, hardness, etc.	Spec(s).
Spline	—A gear-like coupling mechanism for connecting shafts, transmitting mechanical power, and attaching parts.	
Spot Face	—The machining of a circular area around a hole in a weldment or casting to provide a flat, smooth bearing to accept the head of a fastener.	Spt. Fc.
Square	—The plane geometric figure having four equal sides and four interior 90 deg angles.	Sq.
Standard	—Data established by authority, custom, or general consent as a rule for measuring quantity, weight, extent, value, quality, etc. Usually written and published by such authorities as technical societies, professional organizations, or trade associations.	Stnd.
Stock	—The products of metal manufacturers and processors stored in a machine shop for use in making an object or part. The material that a part is made from.	
Surface	—An exterior or interior area of an object. May be flat or curved.	Surf.

Term	*Definition*	*Abbrev.*
Tangent	—A trigonometric function showing the relationship beween the value of an acute angle in a right triangle and the side opposite that angle and the side adjacent. $\text{Tan } A = \frac{\text{Leg opposite } A}{\text{Leg adjacent } A}$	Tan
Tap	—A cutting tool used to form internal screw threads.	
Temperature	—The degree of hotness or coldness of a substance or object measured in degrees Fahrenheit or Celsius.	Temp. (°F or °C)
Tensile Strength	—The maximum stress that a material subjected to a stretching load can withstand without tearing. Also called *hot strength.*	Ten. Str.
Thread(s) (ing)	—The interior or exterior continuous helical rib on a screw or pipe used to join and hold parts together. To cut screw threads in a material.	Thd.
Thru Hole	—The depth of a hole made in a part that goes "all the way through."	Thru
Total Indicator Reading	—A maximum/minimum reading obtained from a full sweep of a part with a dial indicator. Dial indicators are used to check linear alignment, circular positioning, and measure of travel, etc.	T.I.R.
Turn	—The cutting operation performed on the lathe.	
Undercut (ting)	—An underside recess, either cut or molded into an object, that leaves a topside lip or protuberance. To turn a section of a shaft undersize to provide clearance; to bore out a shoulder for clearance.	
Unified National Coarse	—The unified United States, United Kingdom, and Canadian screw thread form having a National Coarse designation.	UNC
Unified National Fine	—The unified United States, United Kingdom, and Canadian screw thread form having a National Fine designation.	UNF
Unified National Special	—The unified United States, United Kingdom, and Canadian screw thread form having a special designation.	UNS
Unified Thread Form	—The United States, United Kingdom, and Canadian screw thread form having a radius at the root instead of a flat as in the National system. The National and Unified forms are interchangeable.	U

Term	*Definition*	*Abbrev.*
Vertical	—Perpendicular to the plane of the horizon; upright.	Vert.
Volume	—The measure of the size of an object or defined region in three-dimensional space. The product of height × width × depth, measured in cubic units: cubic feet, cubic inches, cubic centimeters, etc.	Vol.
Web	—A thin metal section between the ribs, bosses, or flanges of a casting to add strength. In a forging, the thin metal section remaining at the bottom of a depression or at the location of the punches.	
Weight	—The gravitational force with which the Earth attracts a body or object measured in pounds, ounces, kilograms, or grams.	Wt.
Yard	—A unit of length used in the United States and United Kingdom equal to 3 ft, 36 in., or 0.9144m.	Yd.

APPENDIX A

Answers to Training Practice, Chapter II

CHAPTER II: ANSWERS TO TRAINING PRACTICE

Part A (Missing Lines)

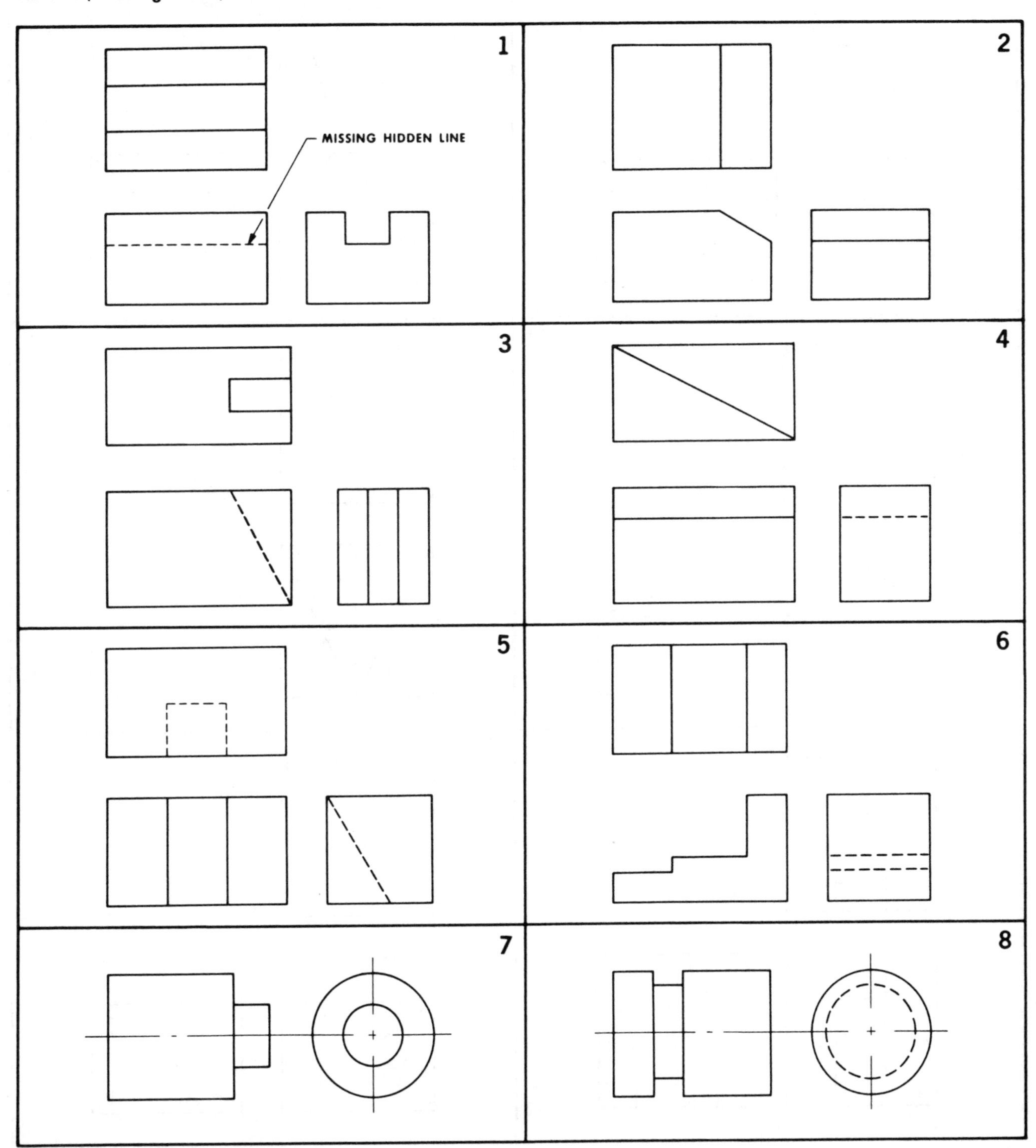

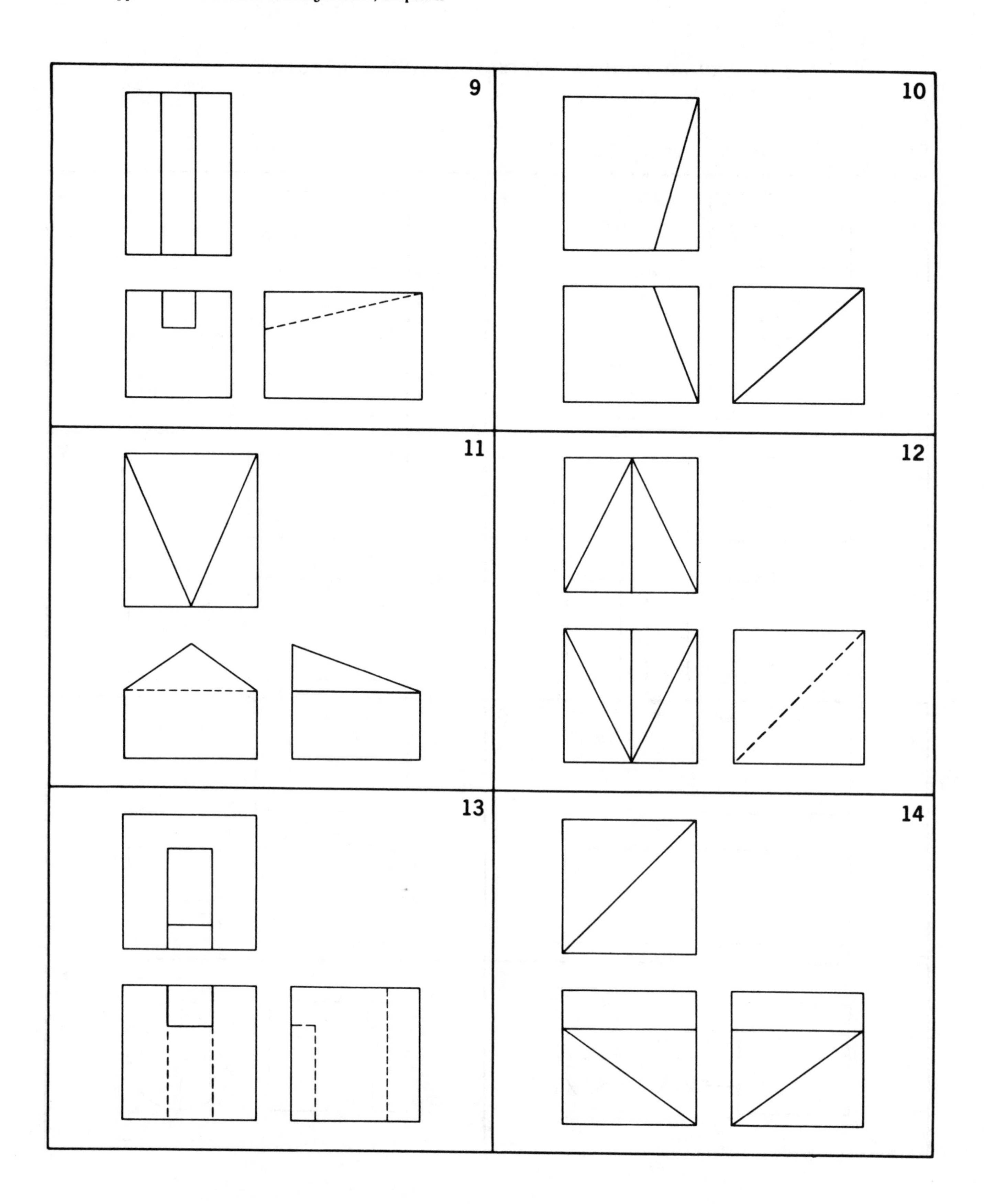
9
10
11
12
13
14

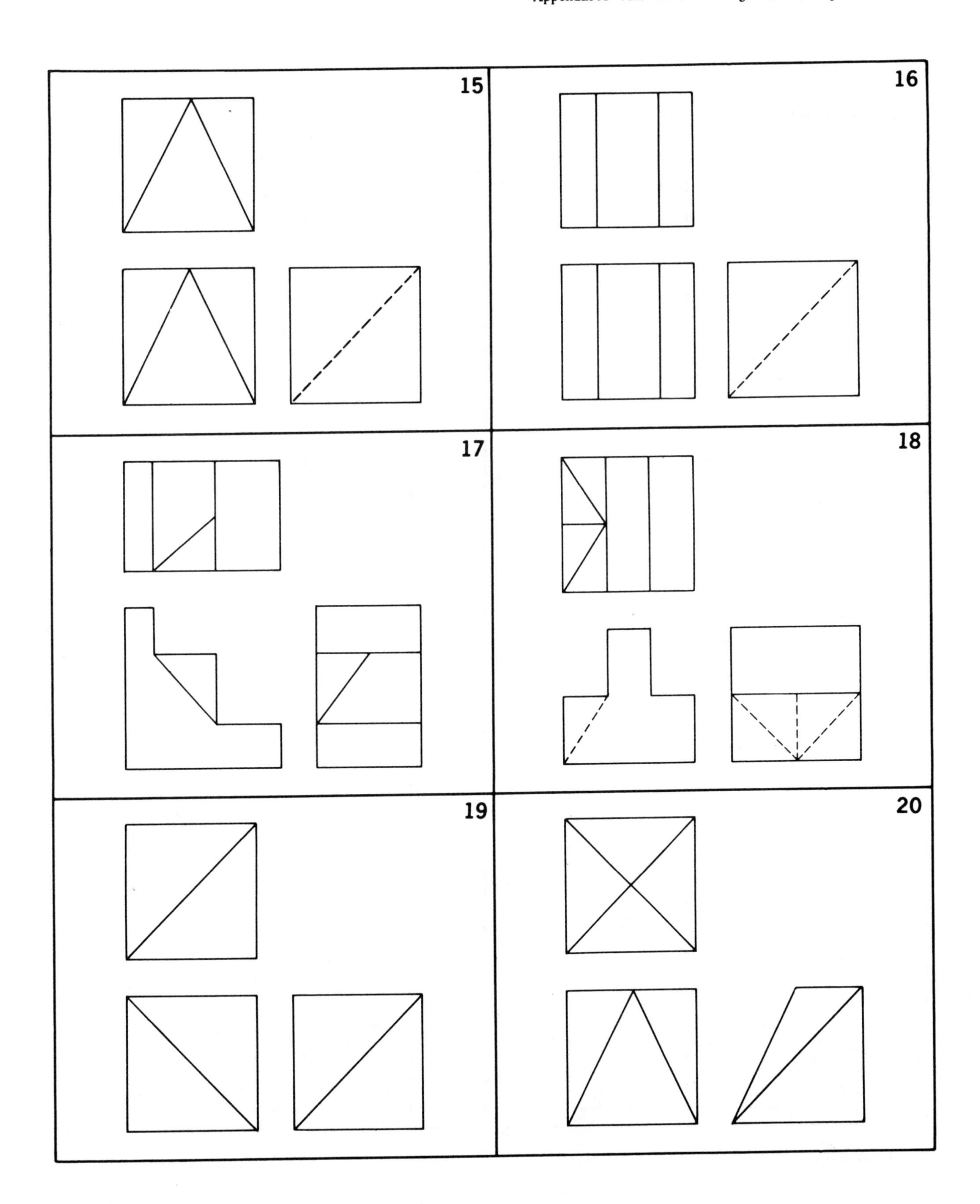
15
16
17
18
19
20

Part B (Missing Views)

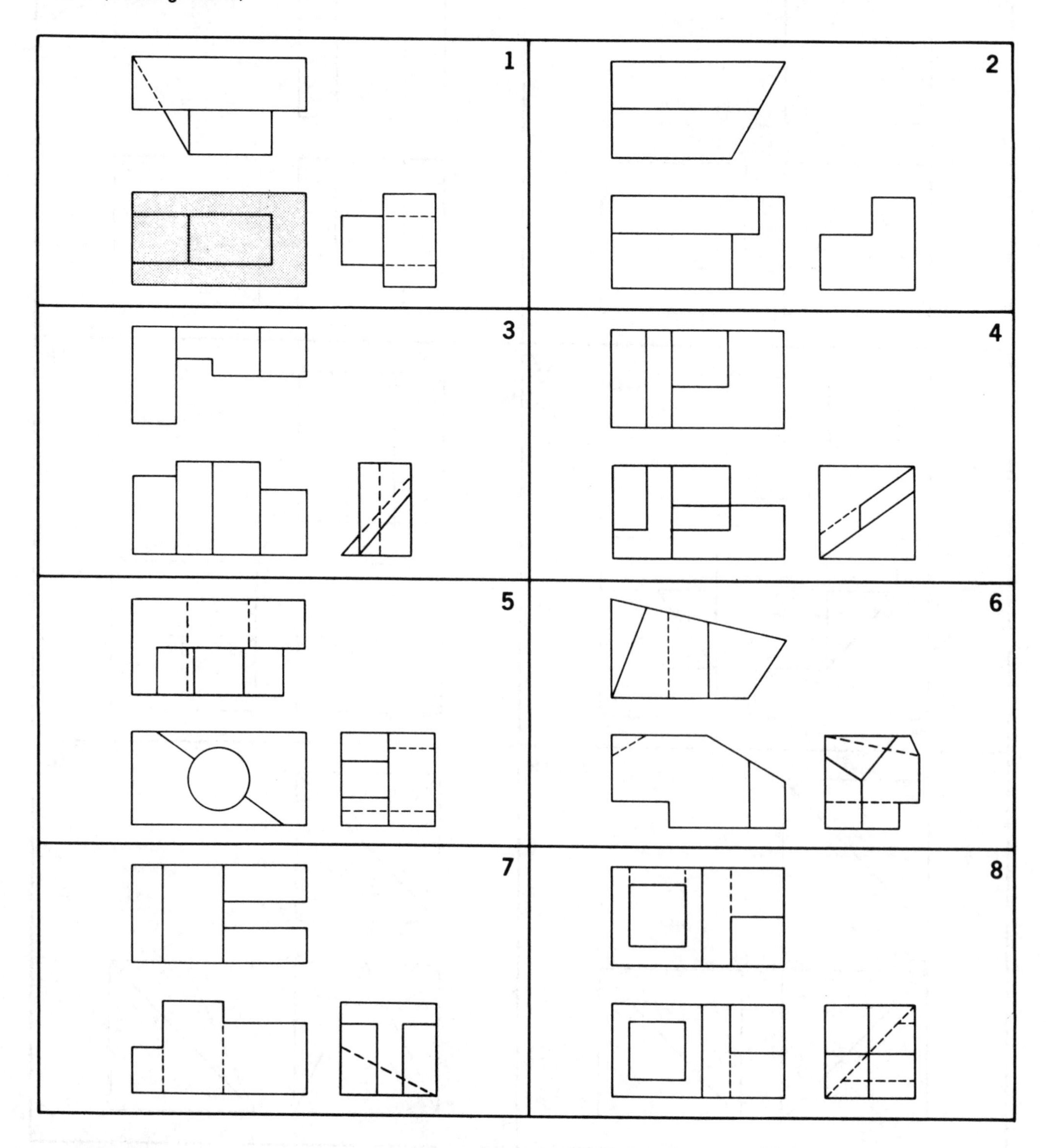

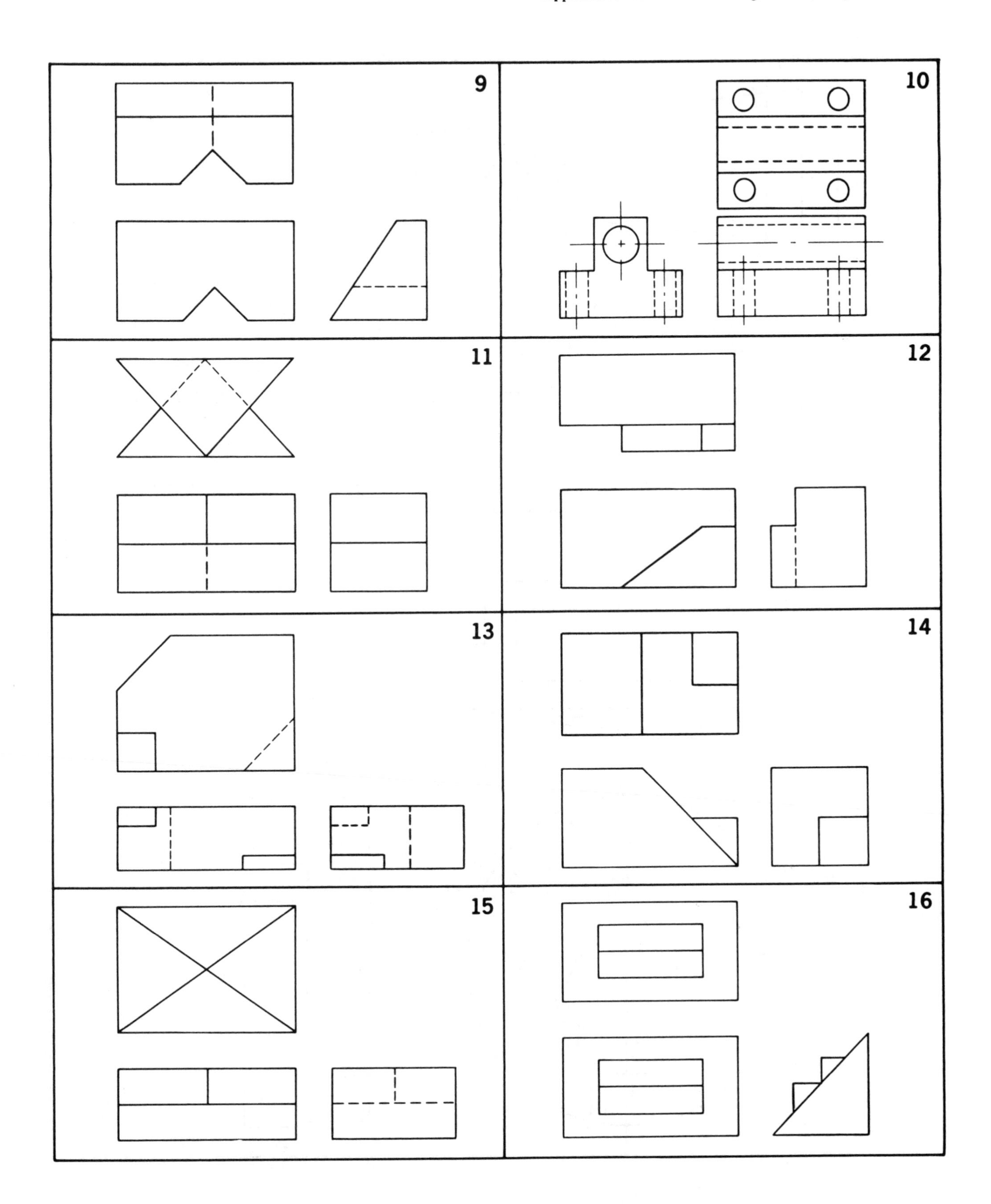
9
10
11
12
13
14
15
16

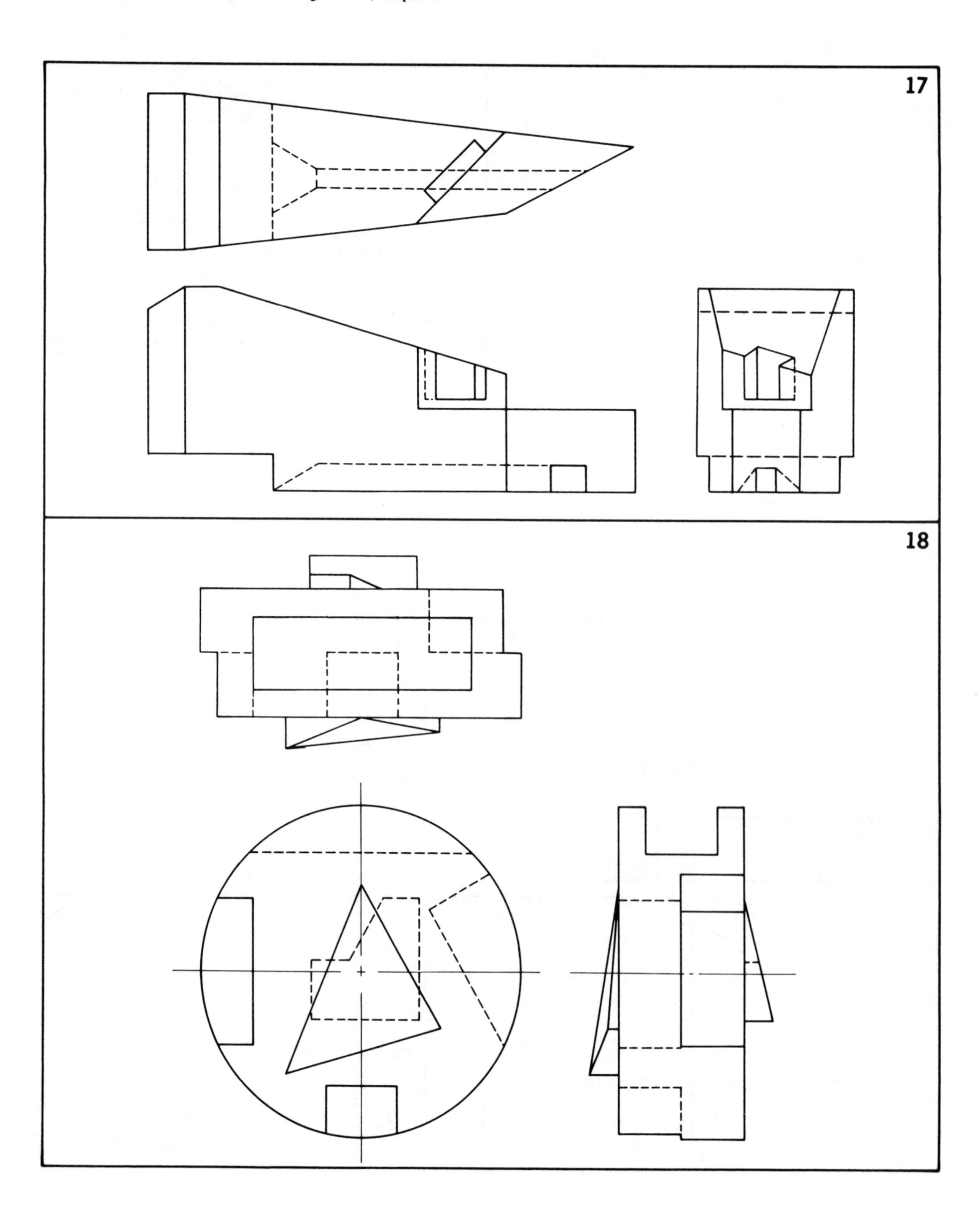
17
18

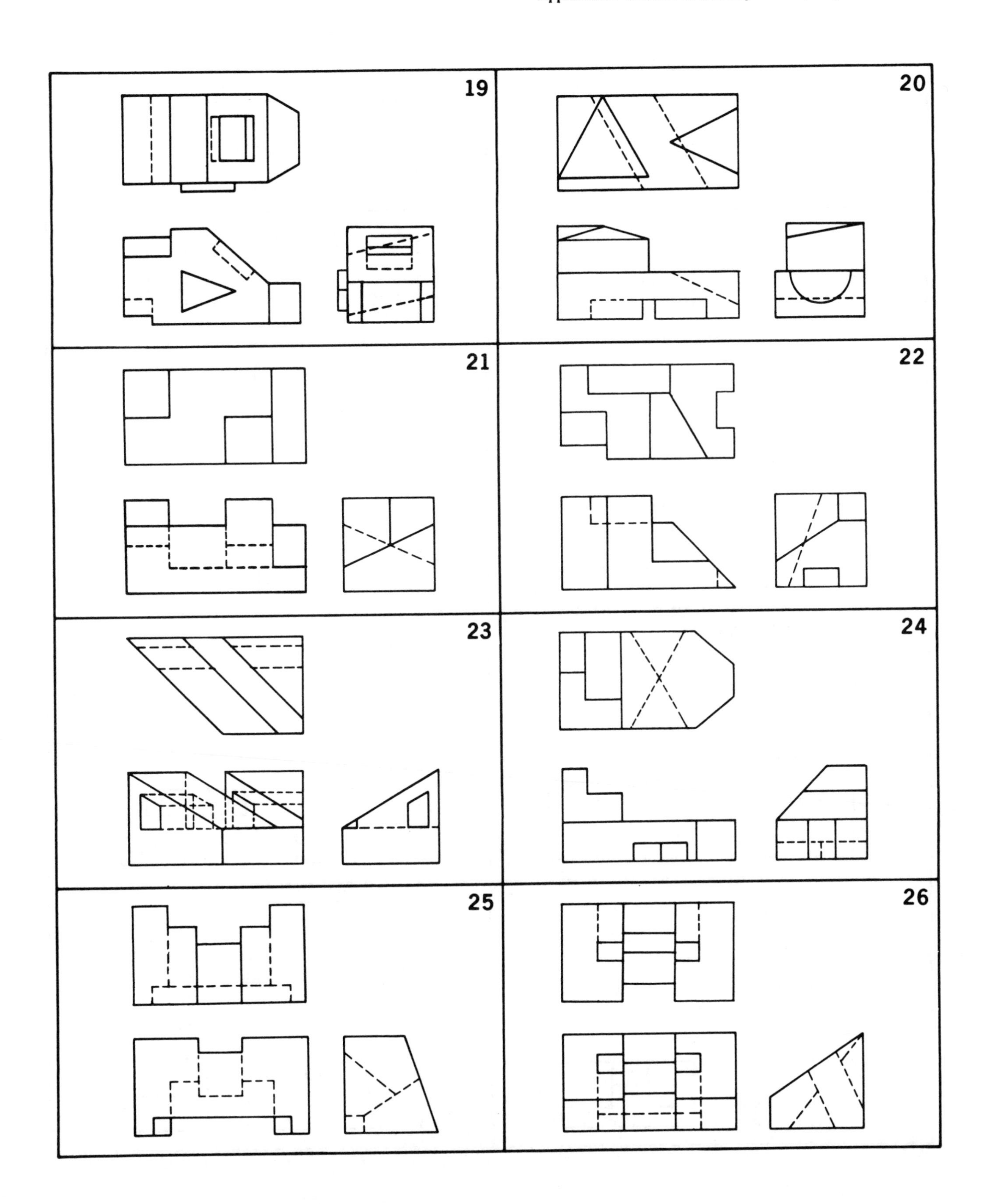
19
20
21
22
23
24
25
26

Part D (Developing an Orthographic Drawing)

1

2

3

4

5

6

7

8

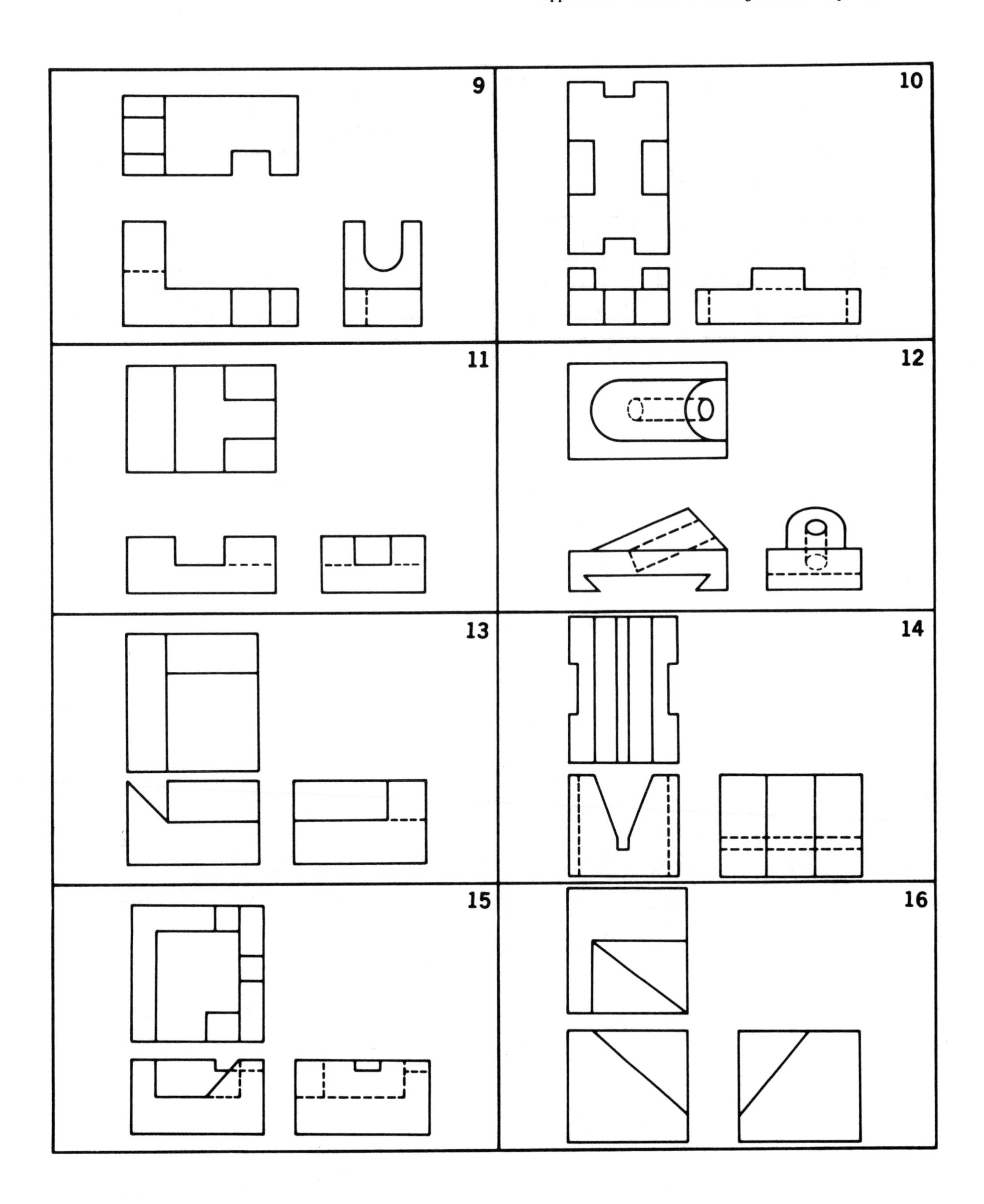
9
10
11
12
13
14
15
16

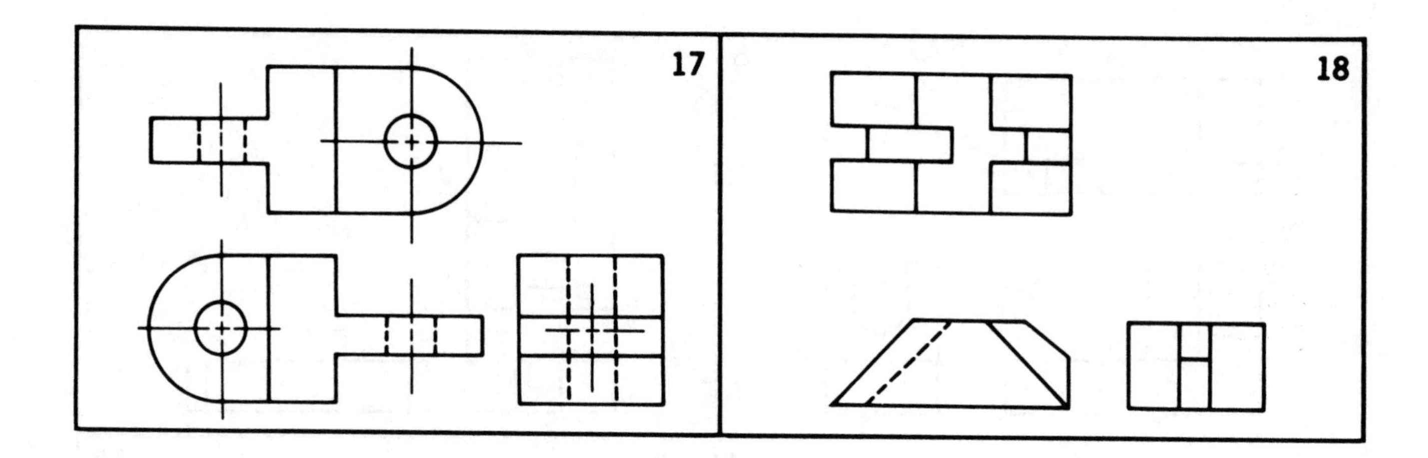
17
18